Learn & Color Nature Series

Garden Edibles

Featuring

seeding square

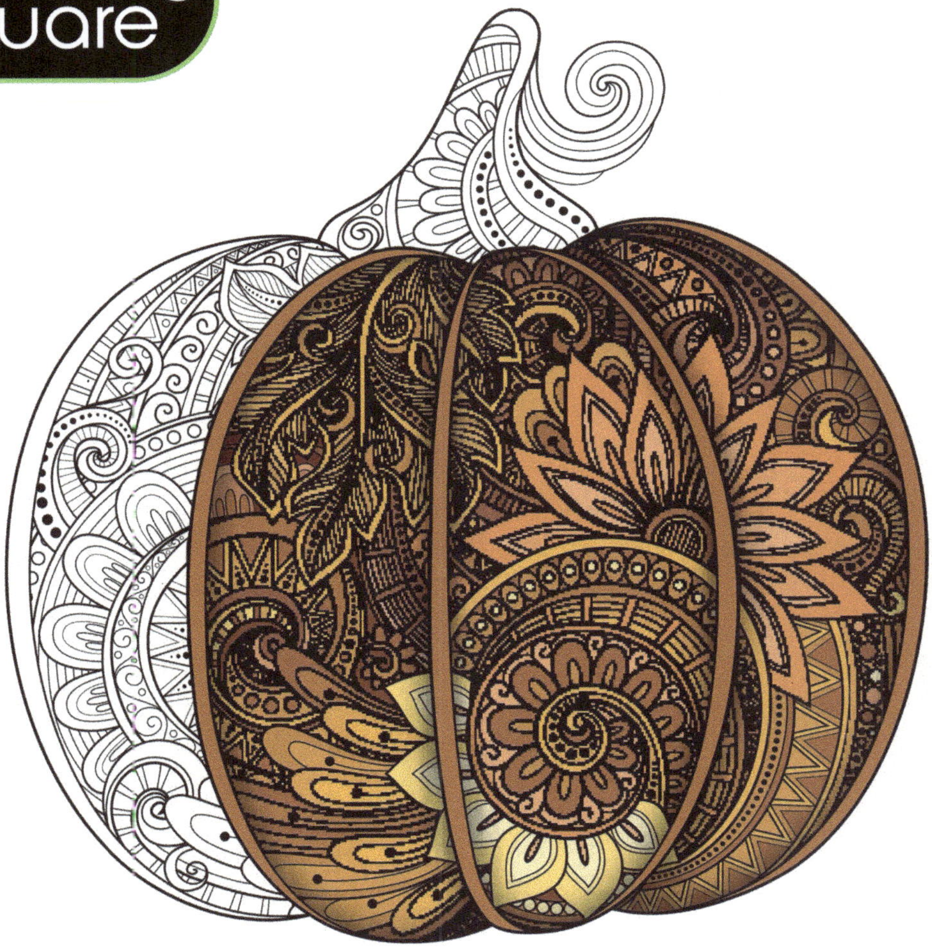

LEARN AND COLOR

Fulton, KY

Subscribe and Save

We offer a new way for you to receive more and more coloring books at a significant savings through our subscription option.

Each month your next coloring book can be delivered right to your mail box. You'll receive a discount from the retail price with shipping and handling included!

Find out all the details and sign up today
http://LearnAndColor.com/educational

Learn and Color Nature Series – Garden Edibles

Learn & Color Books
an imprint of Master Design Marketing, LLC
789 State Route 94 E
Fulton, KY 42041
www.LearnAndColor.com

For information about special discounts available for bulk purchases, sales promotions, fund-raising and educational needs, contact Learn & Color Books Company Sales at sales@LearnAndColor.com.

ISBN: 978-1-947482-12-8

Important. Please Read: This book is for the presentation of vegetables and its contents are for informational purposes only. Never consume any plant without first consulting your doctor. Never take any plant without your doctor's consent if you are pregnant, nursing, or trying to get pregnant. Never give vegetables to children without first talking to their doctor. The statements made in this book have not been evaluated by the FDA. The vegetables mentioned are not intended to cure, treat, diagnose or prevent disease. They make us say this.

Cover and interior design by Faithe F Thomas
Research by Caitlyn F Williams
Photos of Seeding Square © Contour Creations Inc.
Hardiness Maps © USDA (https://planthardiness.ars.usda.gov/PHZMWeb/Downloads.aspx)
All other Photos and Illustrations © 123RF
Text on vegetable pages in this book is a derivative of information by Wikipedia.com, used under CC BY 4.0.
The text of this book is licensed under CC BY 4.0 by Faithe F Thomas.
Look for the Scottish Flag somewhere in each of our books.

seeding square

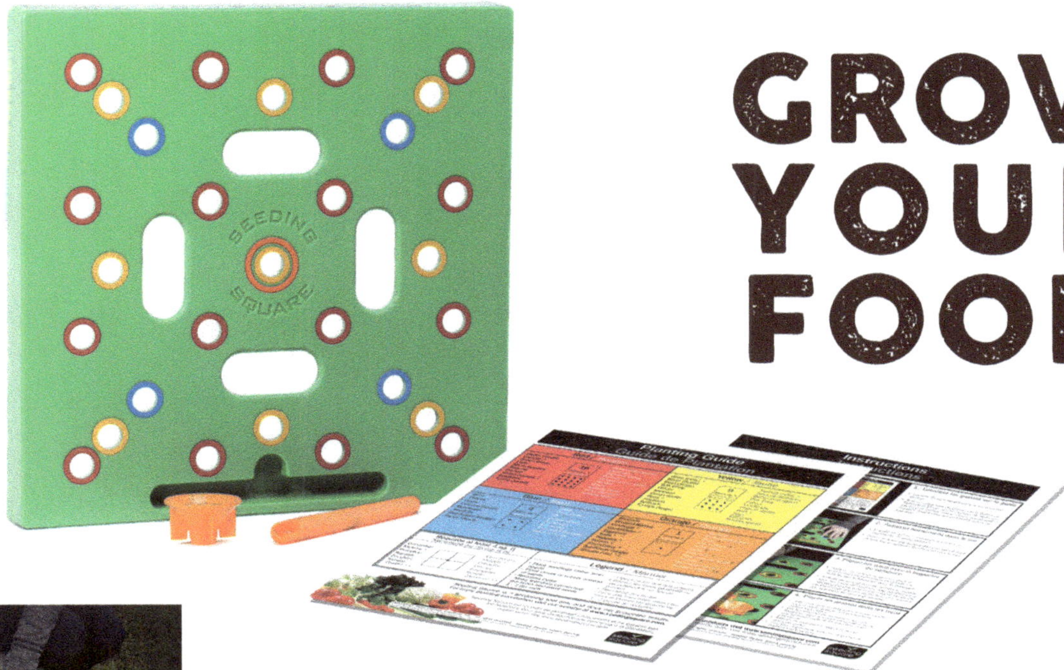

GROW YOUR FOOD™

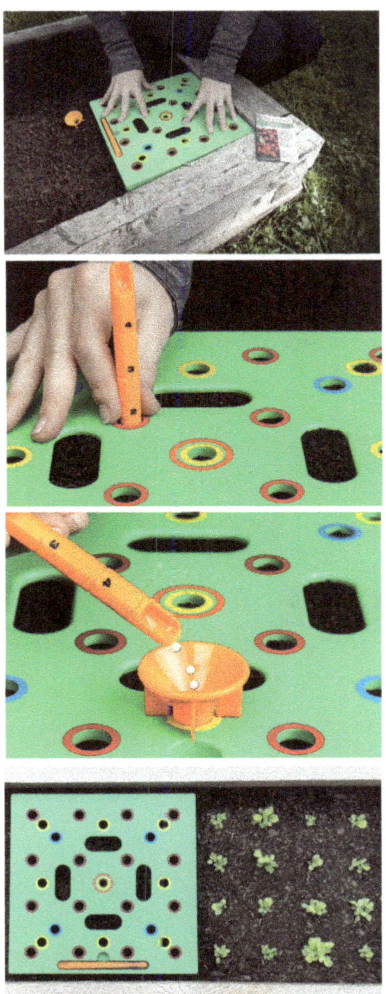

When the Square Foot Gardening method came along, gardening became easier for the garden hobbiest to raise enough food to impact their grocery bill. But it was complicated and time consuming to figure out the grids. And rows were always messy.

So we created a product that makes planting and caring for a vegetable garden simple and fun for all ages and experience levels, while also being on trend for gardening in small spaces. By combining the optimized spacing approach of Square Foot Gardening with the color-coded seeding template, gardens planted with Seeding Square yield a substantial harvest from minimal space. Seeding Square gardeners also find the precision placement of each seed (and seedling) make those pesky weeds easy to spot. Have a junior green-thumb on board? They'll be quick to help on planting day, as the bright colors and poking of holes make planting seeds fun!

It has been proven that gardening is a great stress-reliever. Now, with Seeding Square, it can be accomplished in half the time!

This educational coloring book uses the color coding of the Seeding Square. Look for the easy-to-use planting guide on each page description.

Purchase your Seeding Square at LearnAndColor.com/Square

Asparagus

Asparagus officinalis

Description

Asparagus is a perennial plant that flowers in the spring. Once established, asparagus plants can produce shoots for up to 25 years. It produces individual shoots, which are harvested while still young, and cooked and consumed in various meals. This plant thrives in soil that is too saline for other plants, and rarely has to deal with weeds interfering with the shoots.

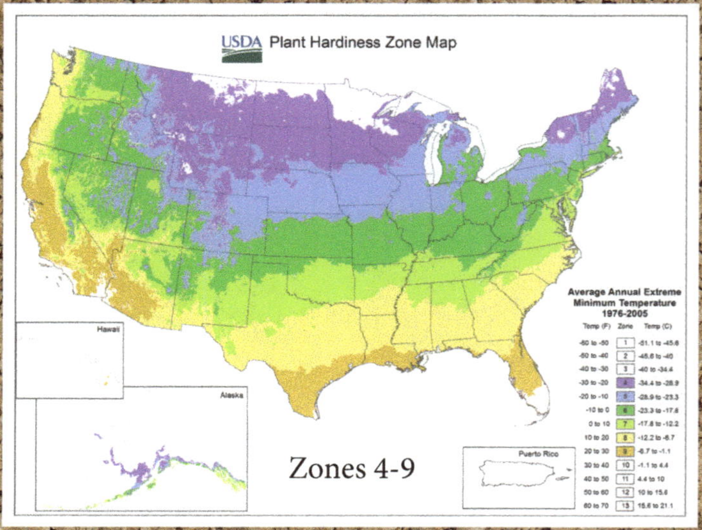

Seeding Square Template

1 plant per square foot, with a minimum of 18" of topsoil (or a raised bed).

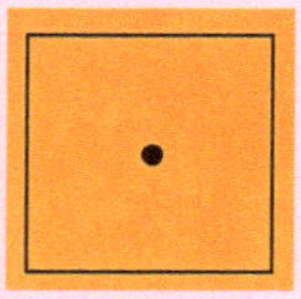

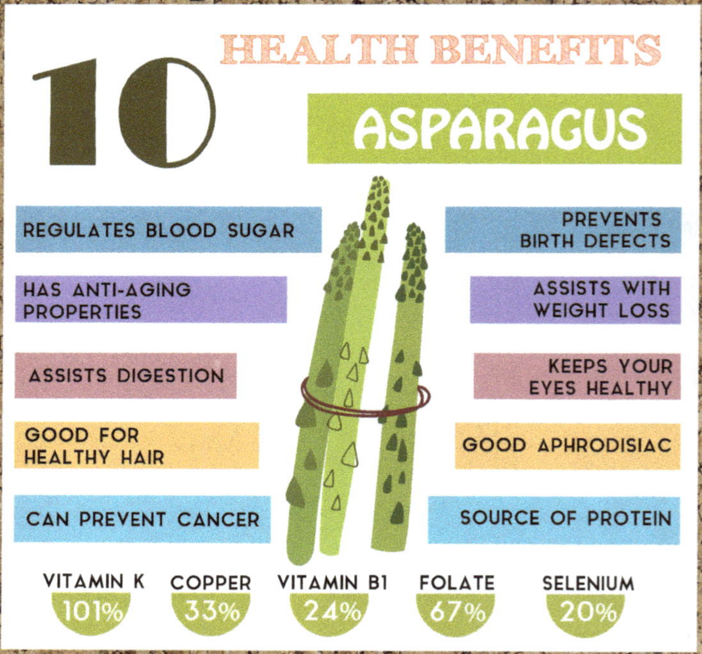

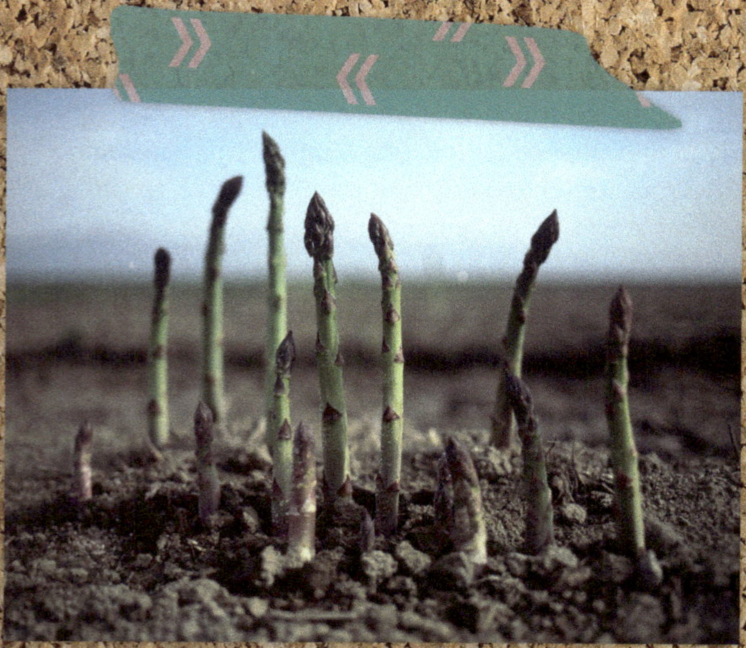

Times to Plant

As soon as the soil can be worked in early spring.

Companion Plants

Basil, Beets, Lettuce, Parsley, Spinach, Tomatoes

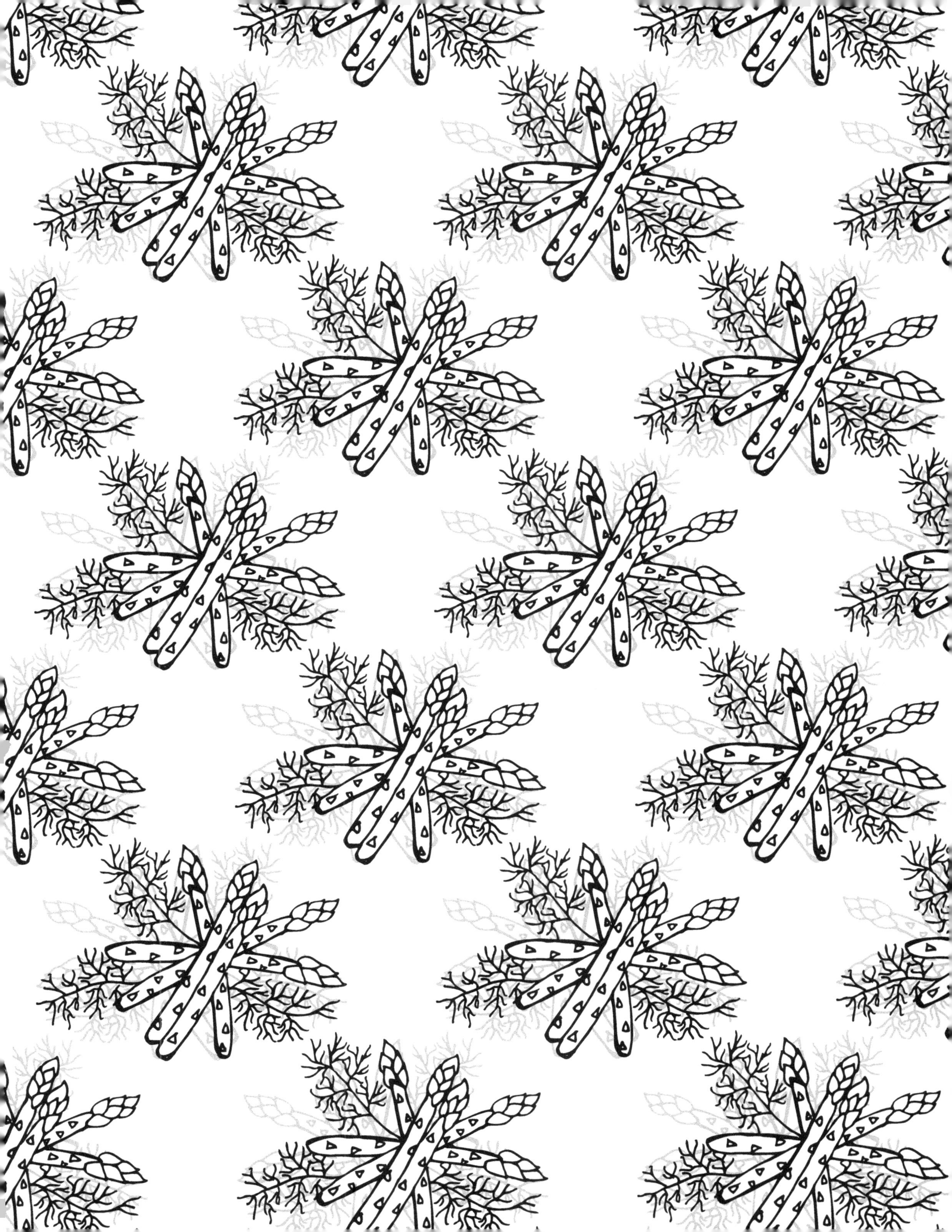

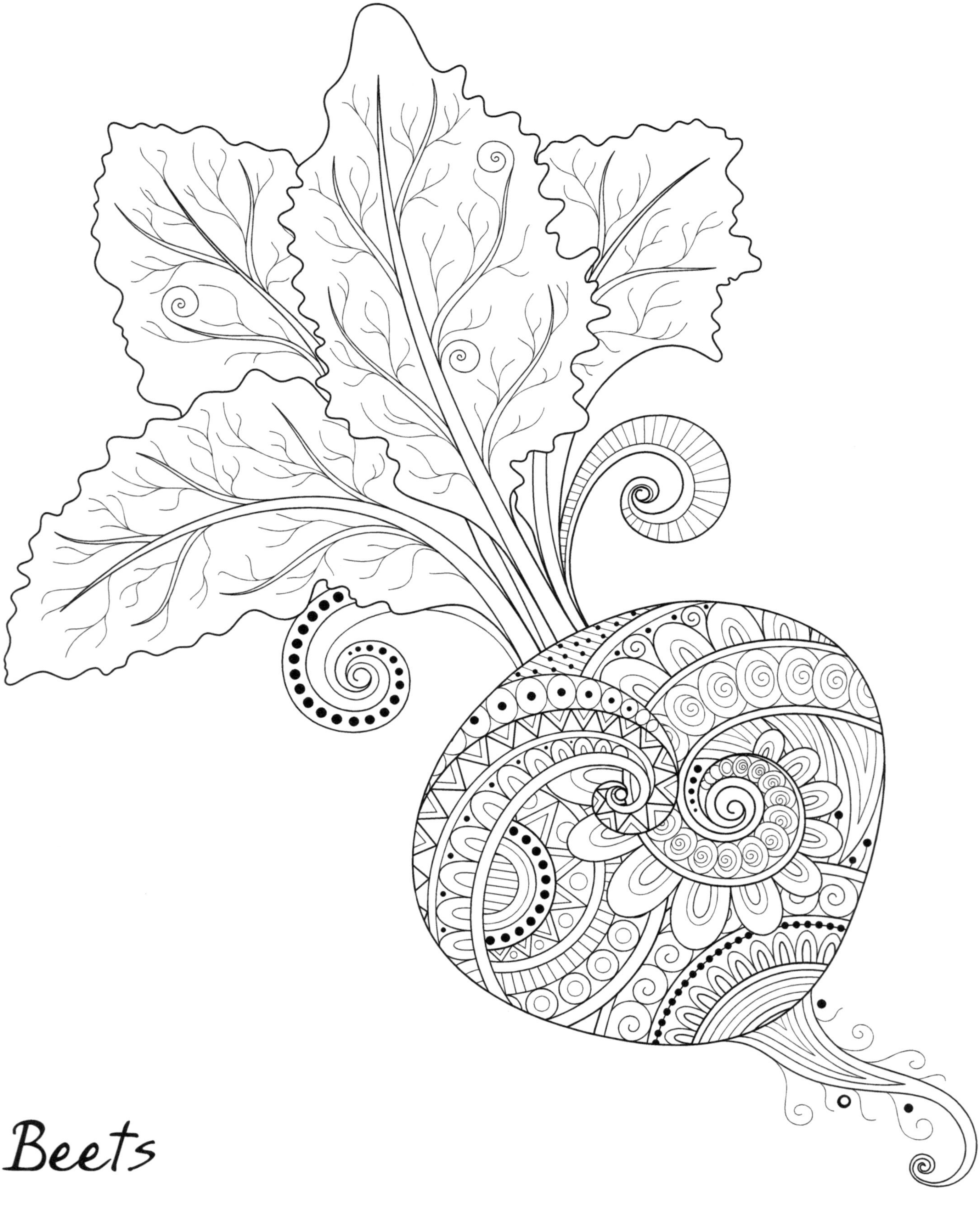

Beets

Beets

Beta vulgaris

Description

Like most other root vegetables, beets are a hearty root that can withstand most chilled temperatures, and can even survive frost. The taproot part of the plant is called the Beetroot. The beet greens are also edible. Other than as a food, beets have been used as a food coloring and as a medicinal plant. They are a dark purple color, and can even be used to make wine.

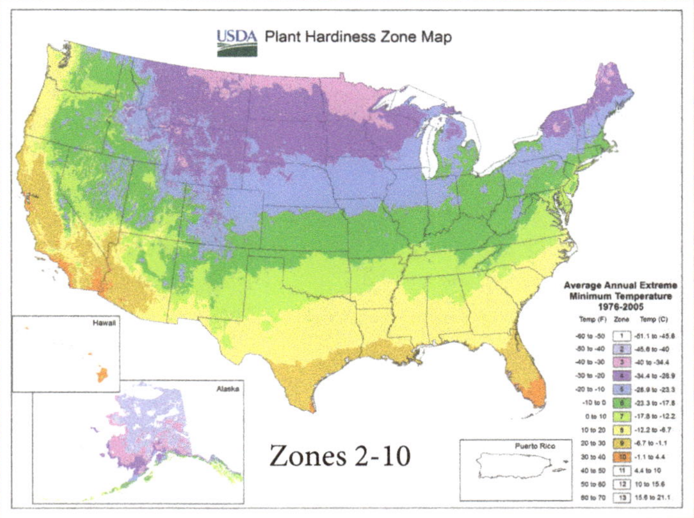

Seeding Square Template

4" per beet, so 9 plants per square foot

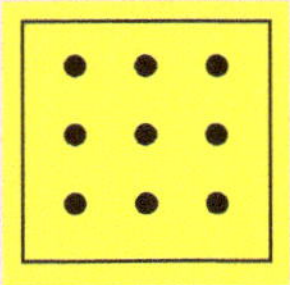

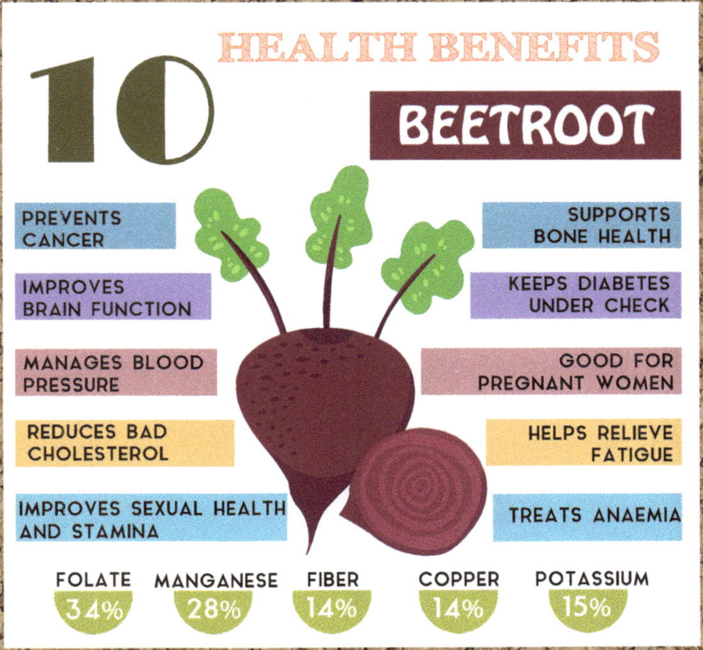

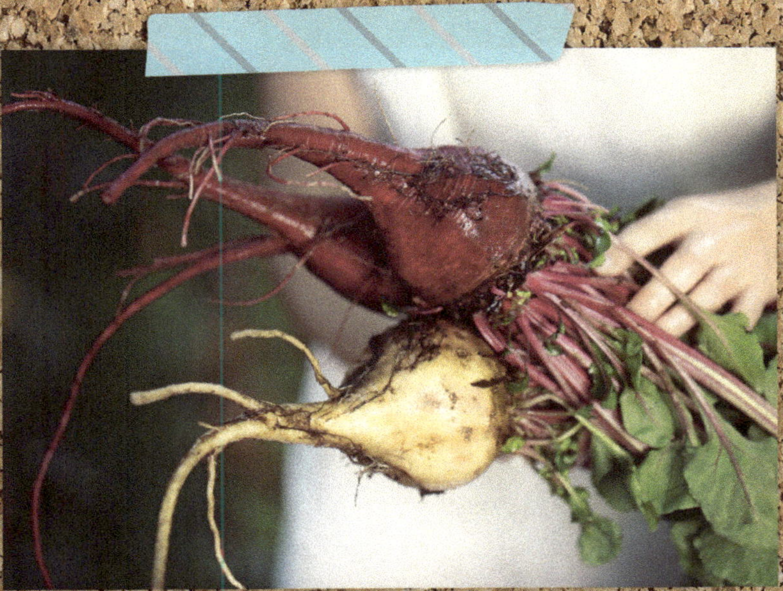

Times to Plant

Beets are hardy enough to be grown all year long, however, they shouldn't be planted until the temperature of your soil is 40° minimum.

Companion Plants

Onion, kohlrabi, bush beans, lettuce, cabbage family

Bell Pepper

Capsicum annuum

Description

Bell Peppers get their name from their large size and their bell shape. They change colors at different stages of ripening, with all of the peppers starting out as green. Red is the most ripe, and has the most nutrients. There are 4 main colors of Bell Pepper – green, yellow, orange and red; however, there are also brown, white, lavender, and dark purple peppers. These peppers actually taste (and even cost) differently, depending on color. They have no tolerance for cold weather, and only thrive in warm, humid climates.

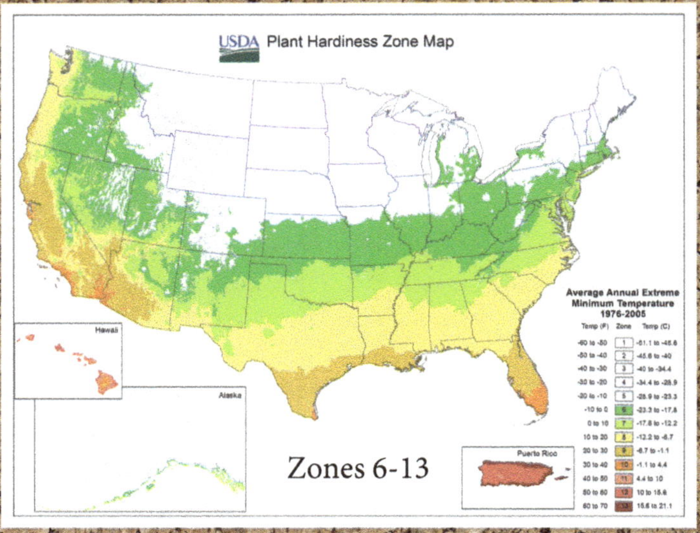

Seeding Square Template

1 plant per square foot

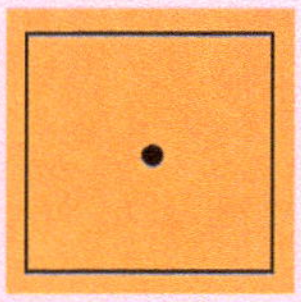

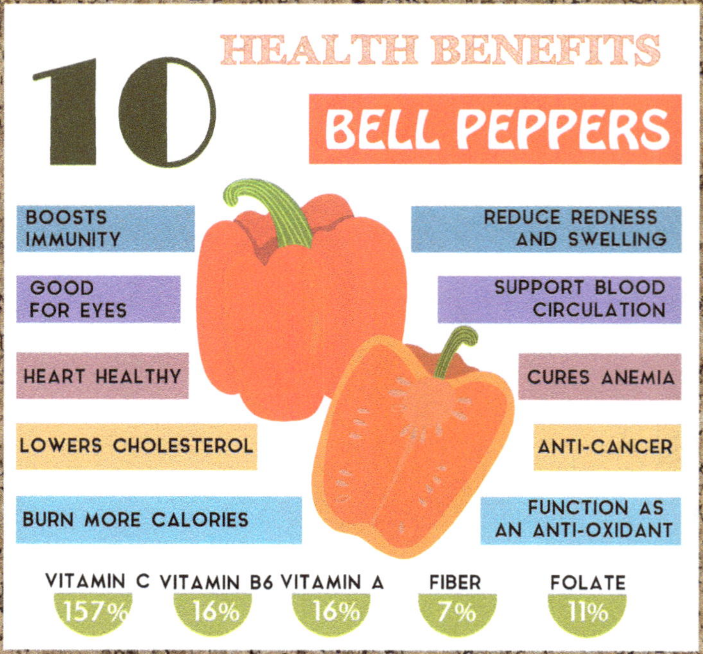

Times to Plant

You should start your seeds indoors, 8-10 weeks before last spring frost date.

Companion Plants

Basil, parsley, oregano, dill, tomatoes, carrots, cucumbers, squash, spinach, and peas.

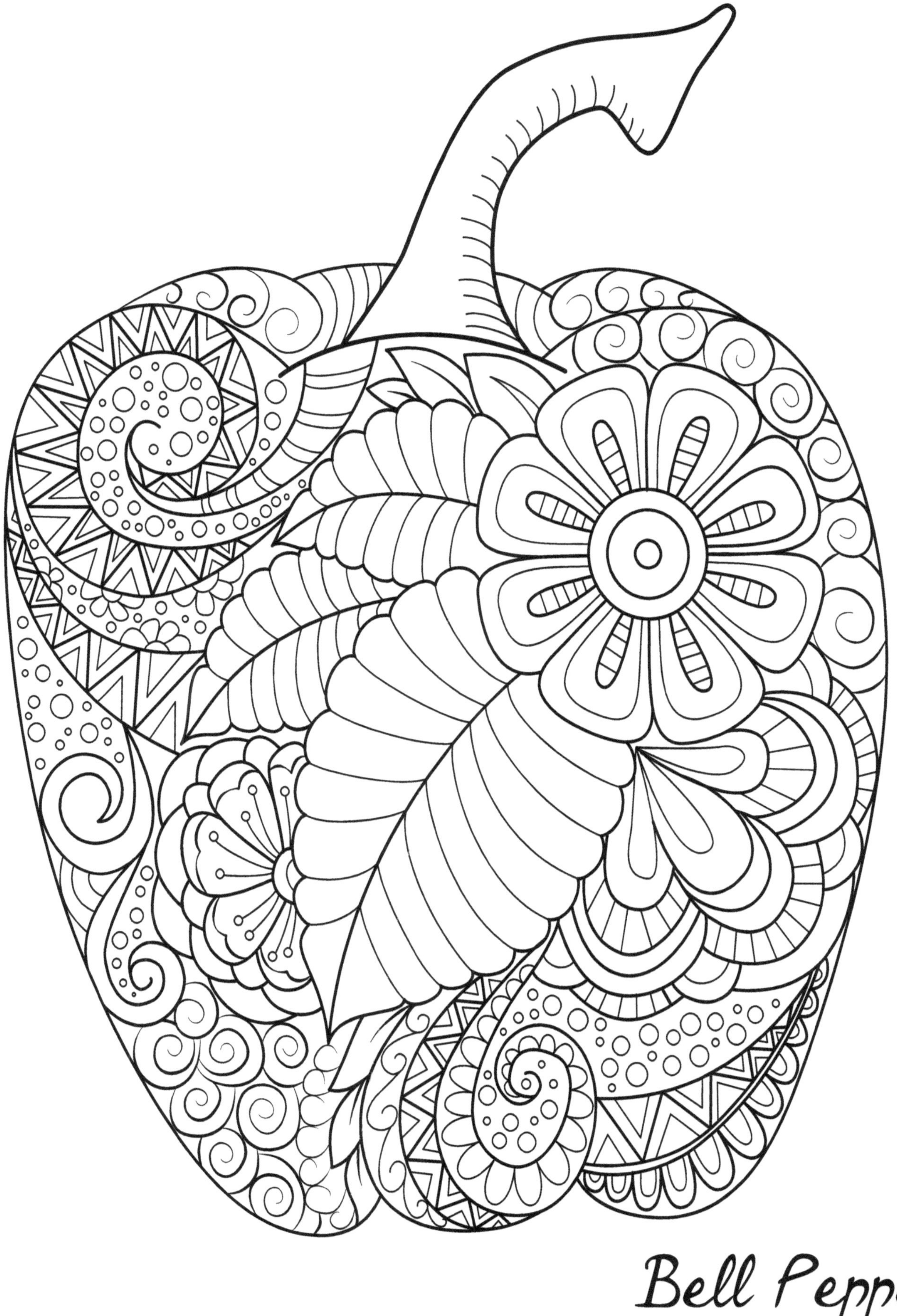

Bell Pepper

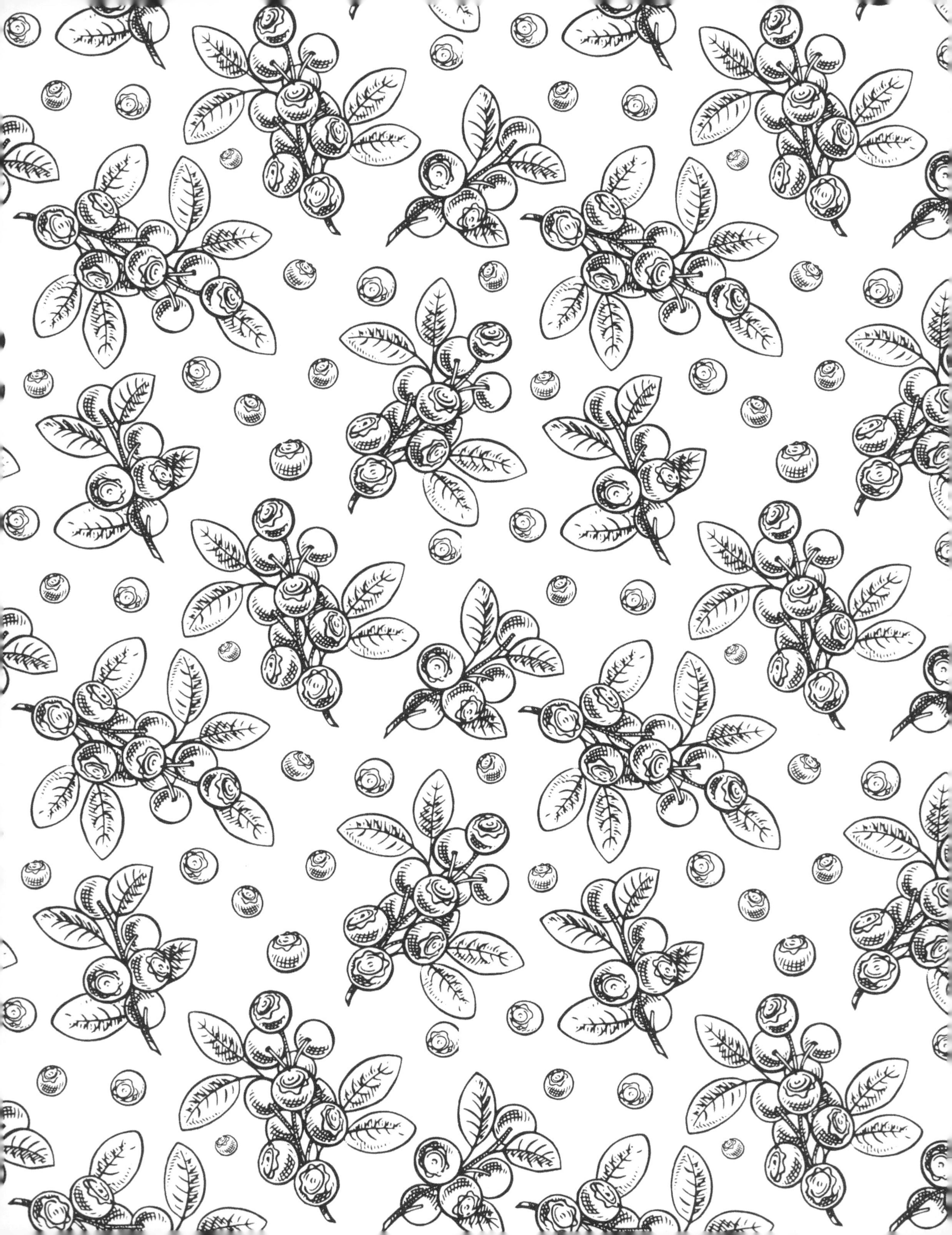

Blueberry

Cyanococcus vaccinium

Description

Blueberries are perennial flowering plants with blue or purple colored berries. Blueberries are usually prostrate shrubs that can vary in size from 4" to 13' in height. There are several types of blueberries, but the most common is the Highbush Blueberry plant. If properly cared for, blueberry bushes can produce berries for up to 20 years.

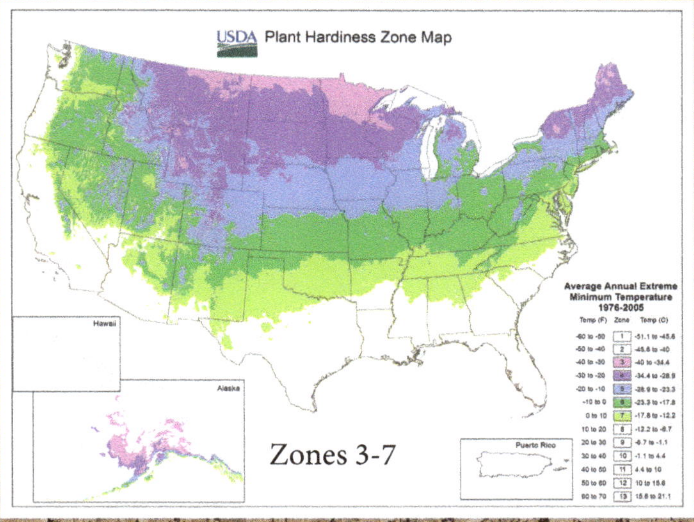

Seeding Square Template

Plant your bushes every 4 to 6 feet

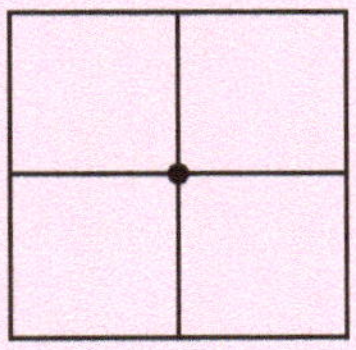

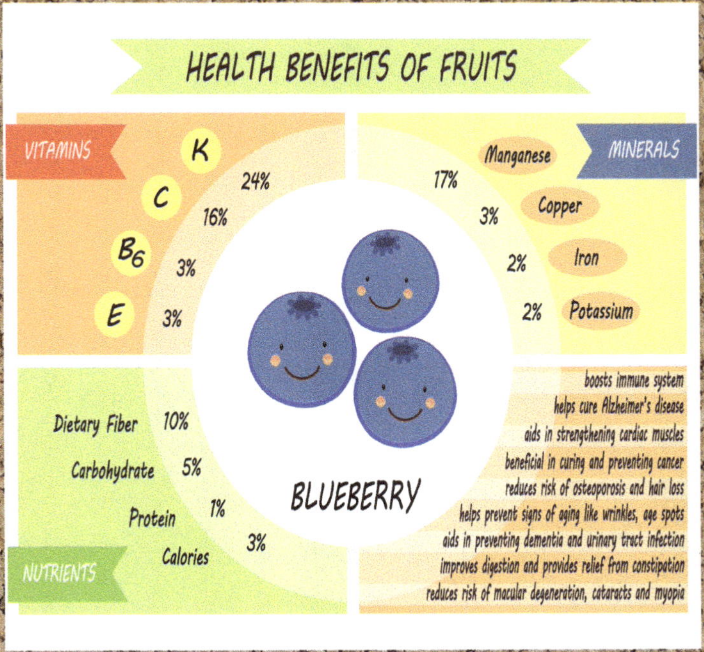

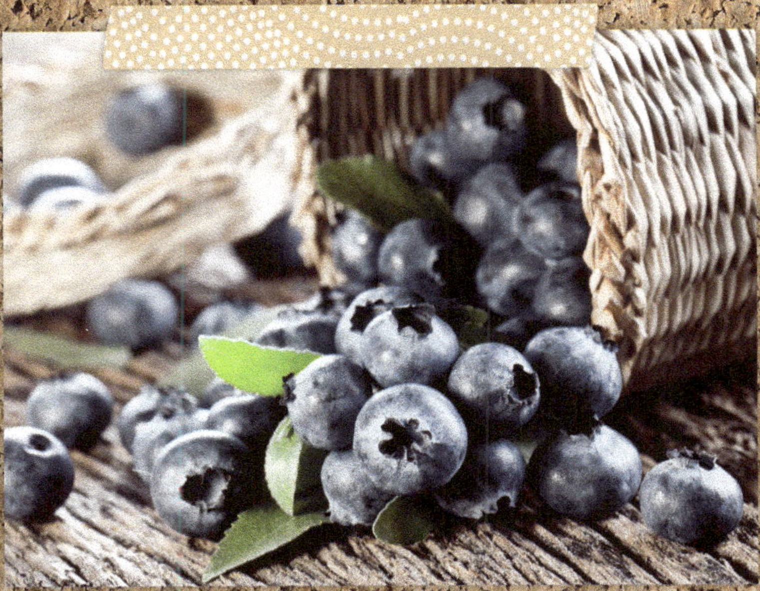

Times to Plant

Early to mid-fall, however, December and January also work well.

Companion Plants

Rhododendron, pine trees, herbs, and ground cover such as grass.

Broccoli

Brassica oleracea

Description

Broccoli is a part of the cabbage family, growing in the cooler months of the year. There are three different types: Calabrese broccoli, which is the most common types, Sprouting broccoli, which has a larger number of heads with many thin stalks, and Purple cauliflower, a type of broccoli that has a head shaped like cauliflower, but consists of tiny flower buds. It sometimes, but not always, has a purple cast to the tips of the flower buds. Cauliflower, kale, and collard greens are also part of the family.

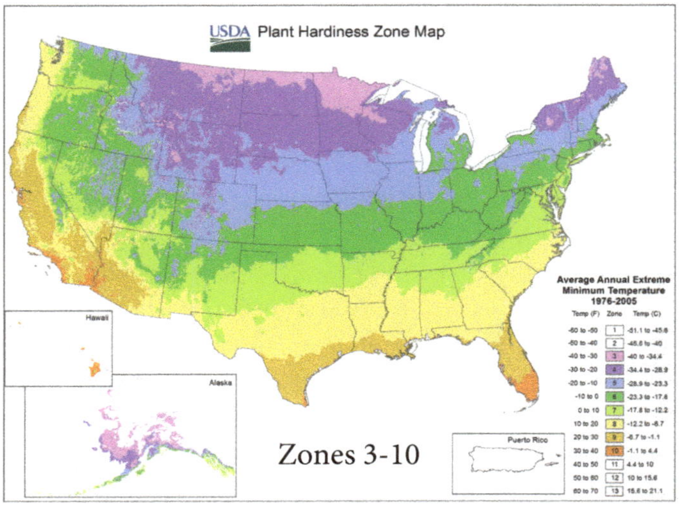

Seeding Square Template

12-18" per plant, with 1 plant per square to 4 plants in 9 squares

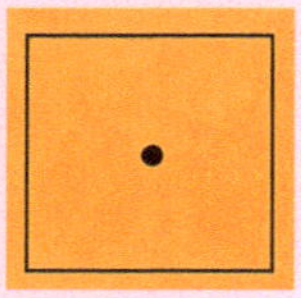

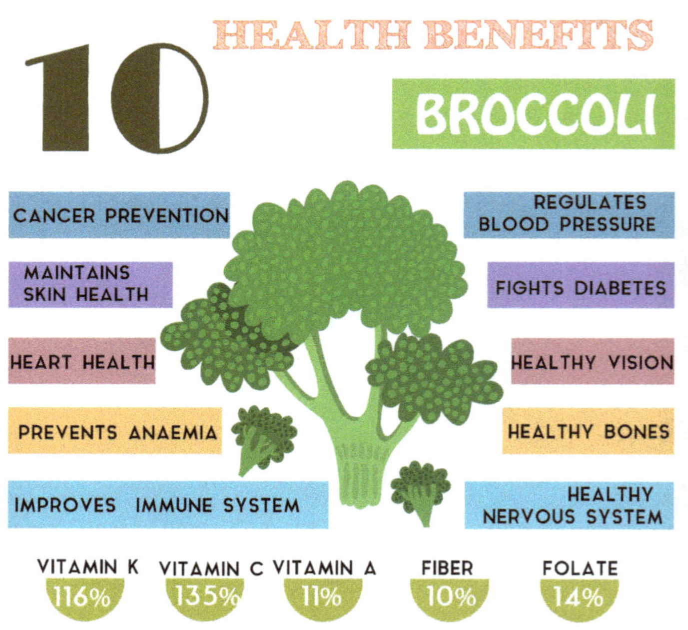

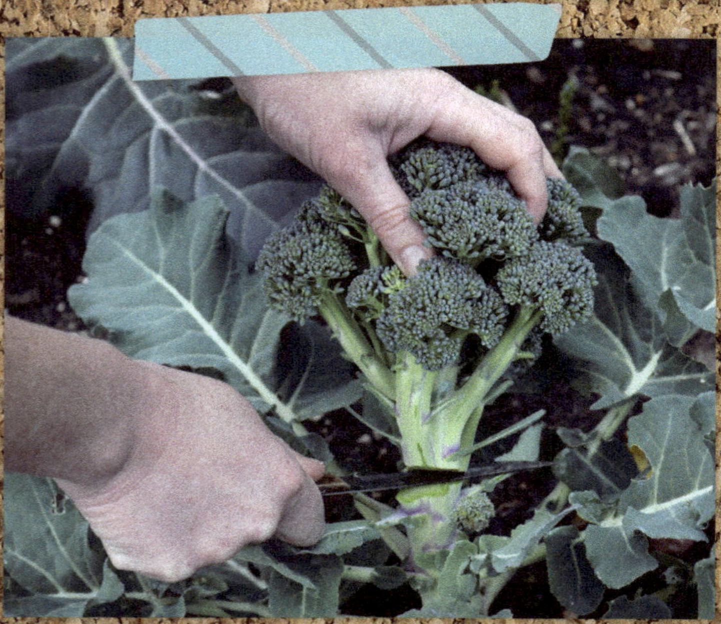

Times to Plant

Broccoli prefers cooler weather, so plant your seeds indoors approximately 12 weeks before your final frost date, and then transplant outside 6 weeks before the final frost.

Companion Plants

Beets, celery, cucumbers, dill, garlic, hyssop, lettuce, mint, onions, potatoes, rosemary, sage, spinach, swiss chard

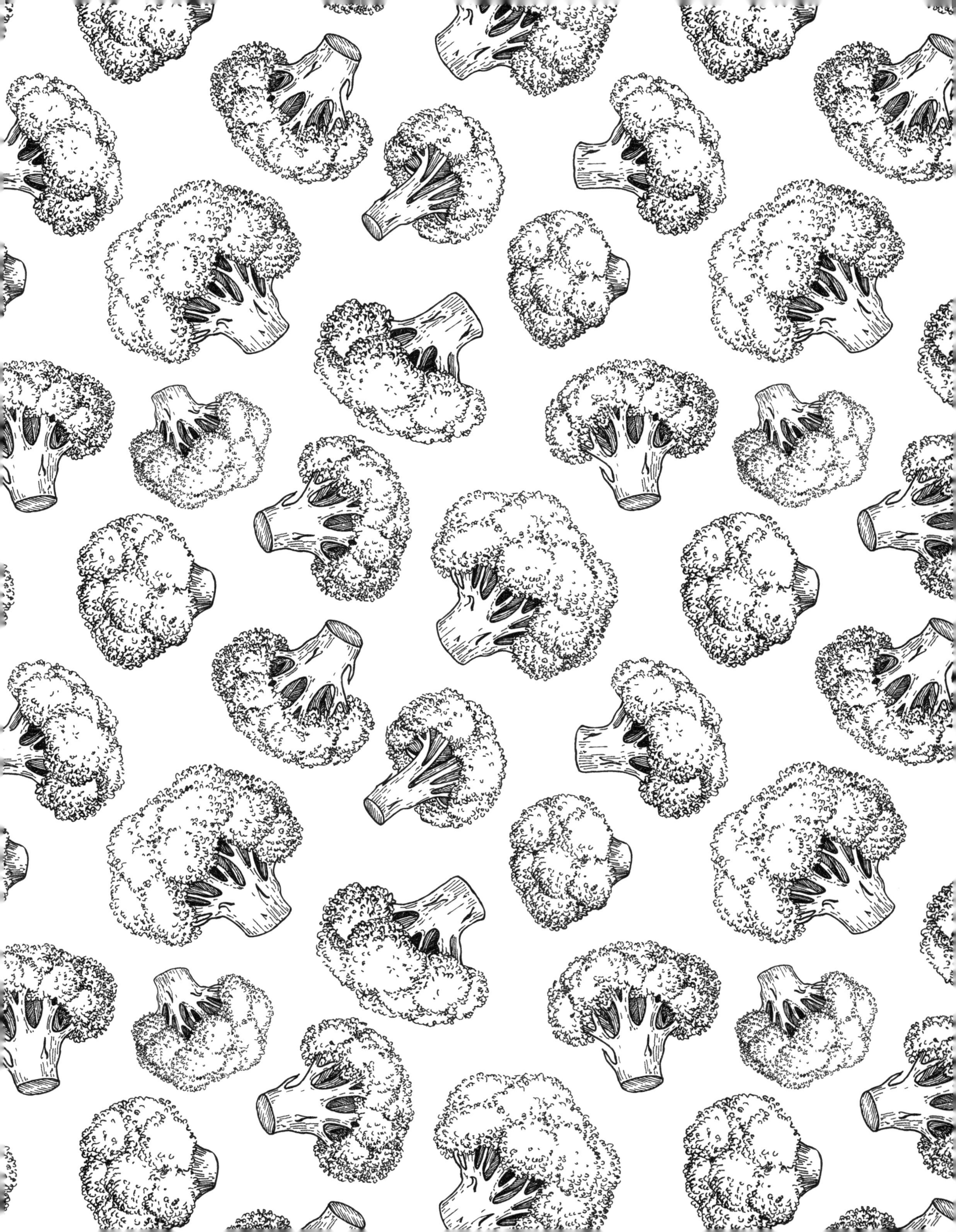

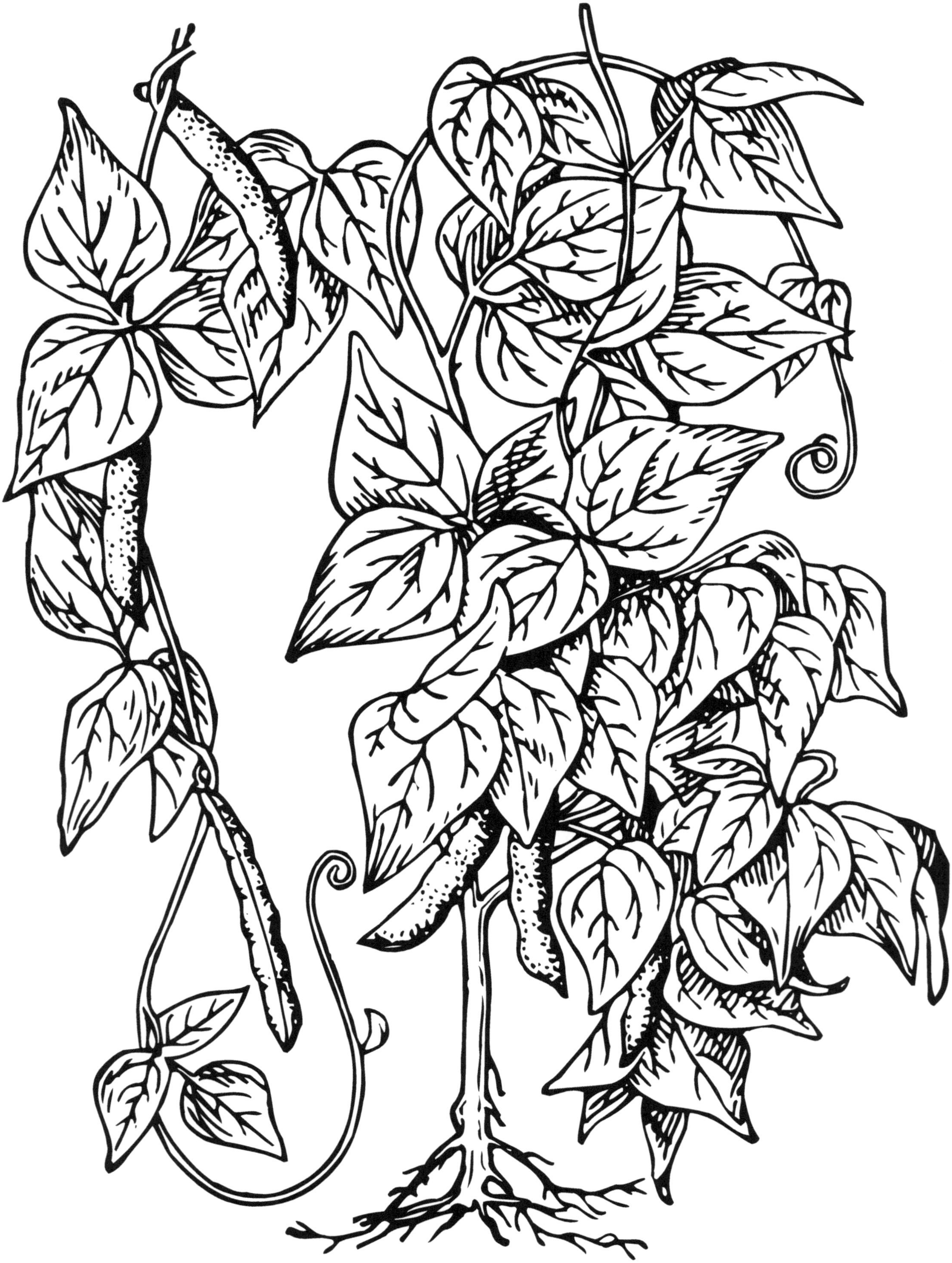

Bush Beans

Phaseolus vulgaris

Description

There are two types of plants when it comes to growing beans. There are bush beans and pole beans. Pole beans need support to grow on, while bush beans support themselves and don't grow long or tall enough to need assistance. And there are 3 different types of bush beans: snap beans, green shelling beans, and dry beans.

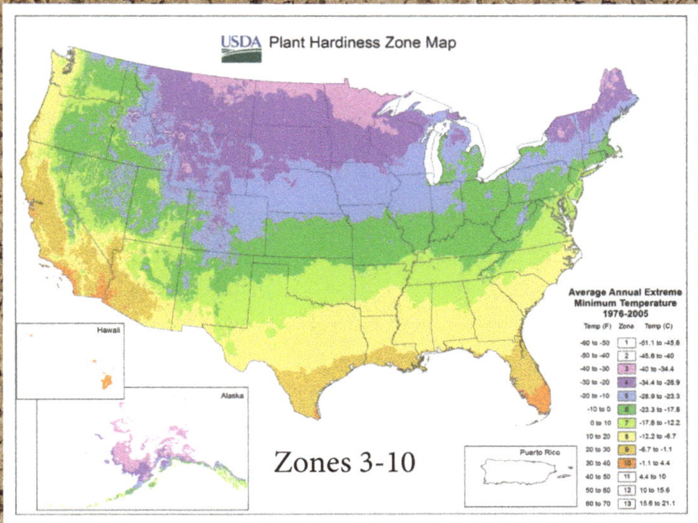

Seeding Square Template

Bush beans seeds need to be 2-3" apart, but will ultimately spread up to 2'.

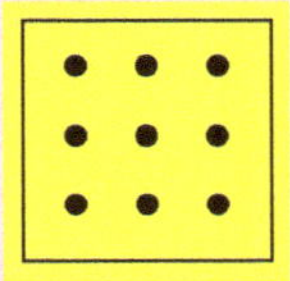

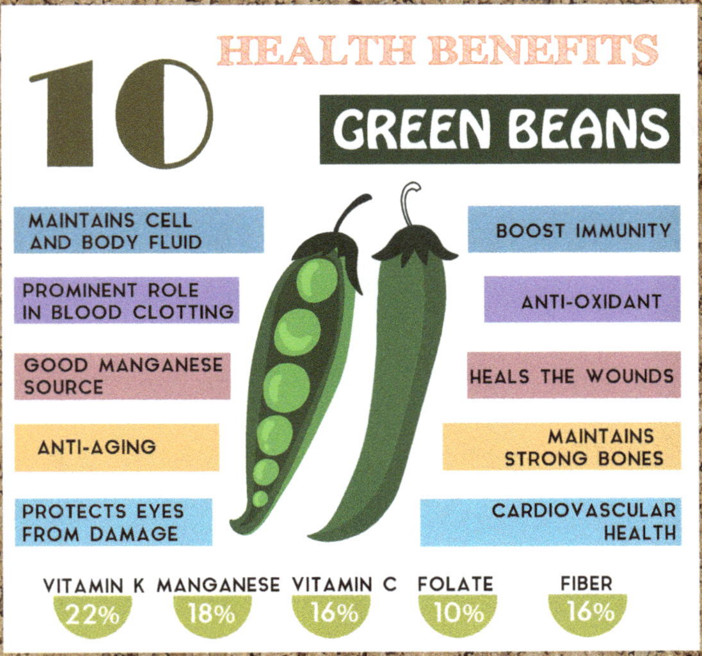

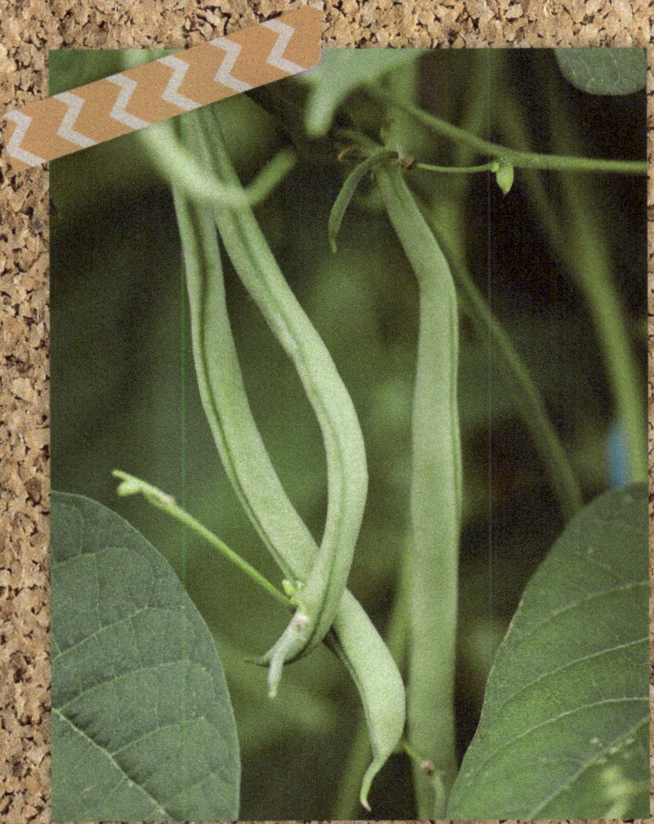

Times to Plant

Seeds can be planted anytime after the last spring frost, as long as the minimum soil temperature is 48°.

Companion Plants

Beets, cauliflower, celery, corn, marigolds, potatoes, strawberries, and sweet potatoes.

Butternut Squash

Cucurbita moschata

Description

Unlike Yellow and Zucchini Squash, Butternut Squash is a winter squash, which means it has a longer growing time, and is harvested once the gourds are mature. These squash grow on vines, as opposed to growing on bushes, and therefore need a bit more growing room. They prefer to grow up and around things, so a trellis will work well for the vining winter squashes.

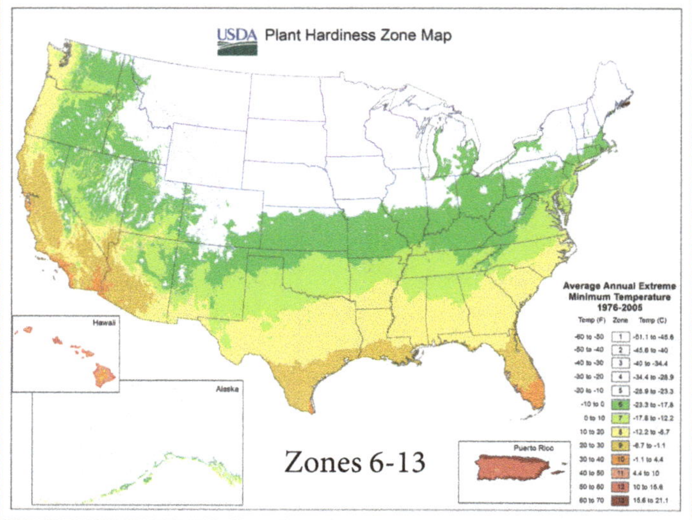

Seeding Square Template

One to two plants per 4 square feet

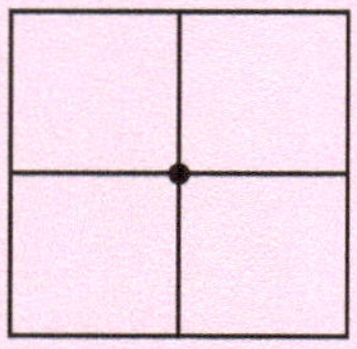

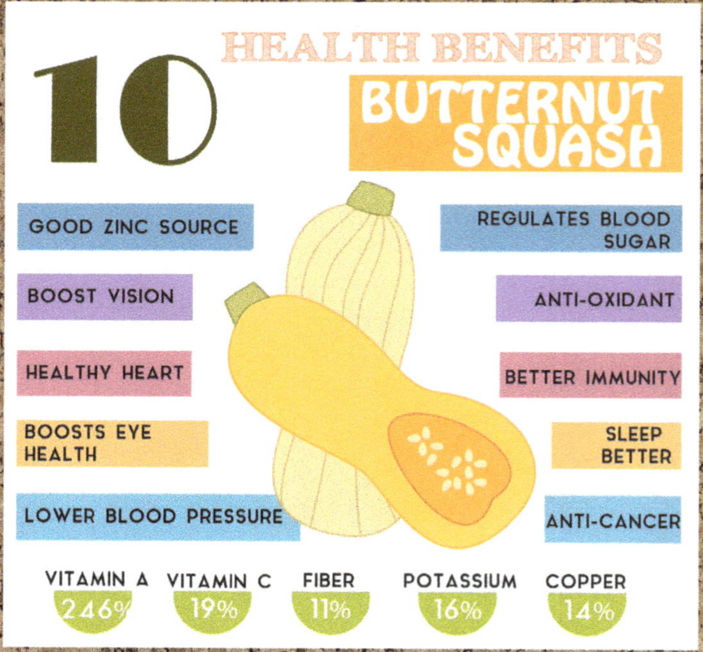

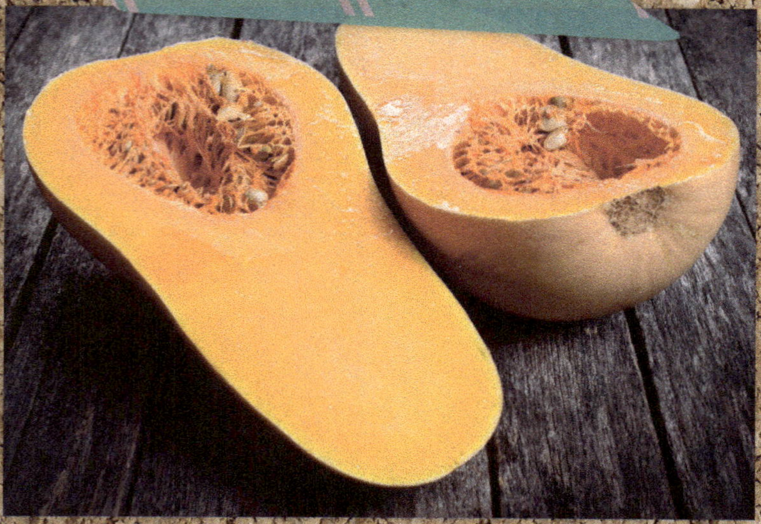

Times to Plant

Plant when you have passed your final frost date, or when outdoor temperatures stay reliably above 70º, and your soil temp is in the 60s.

Companion Plants

Corn, onion, eggplant, marigold, nasturtium, radish

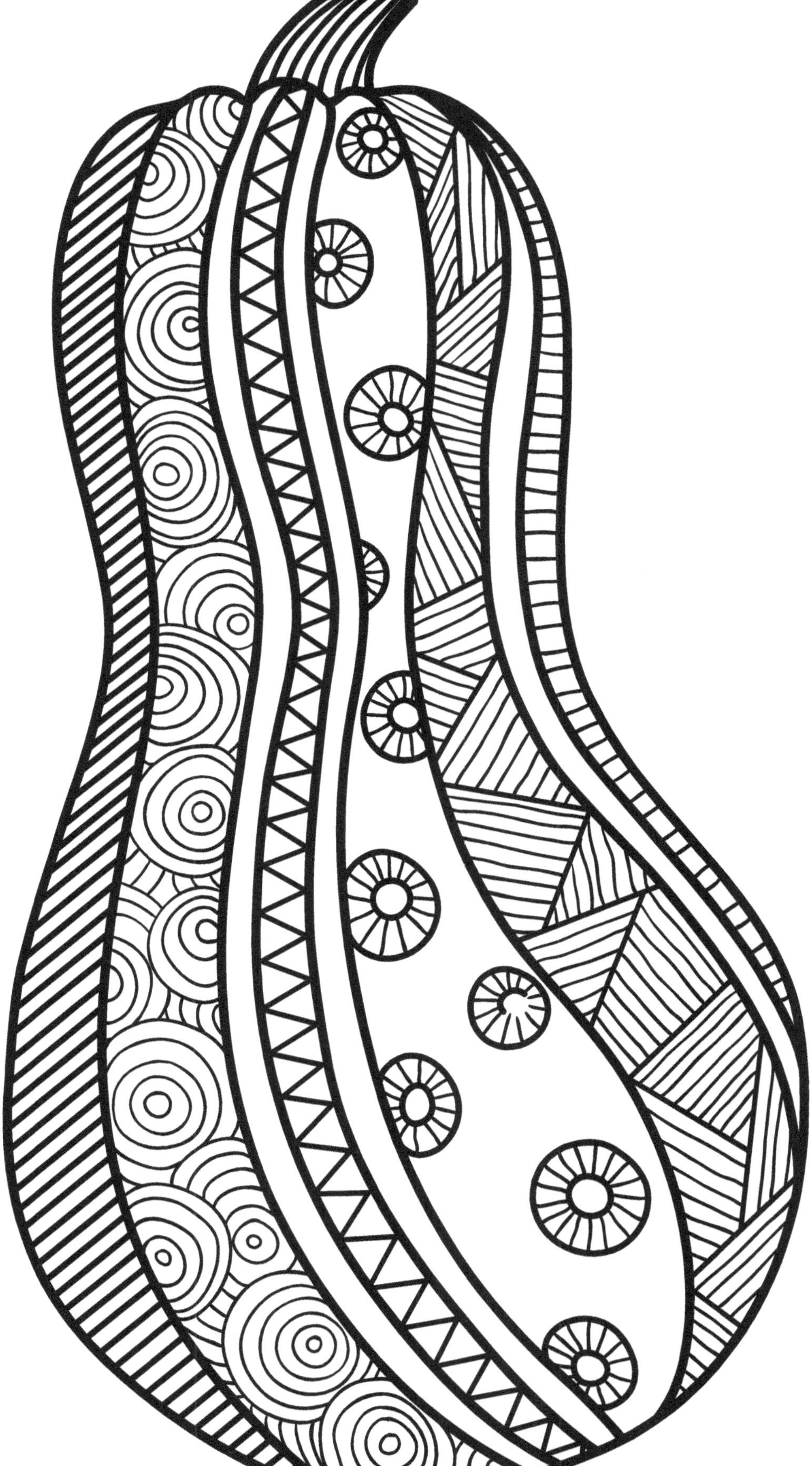

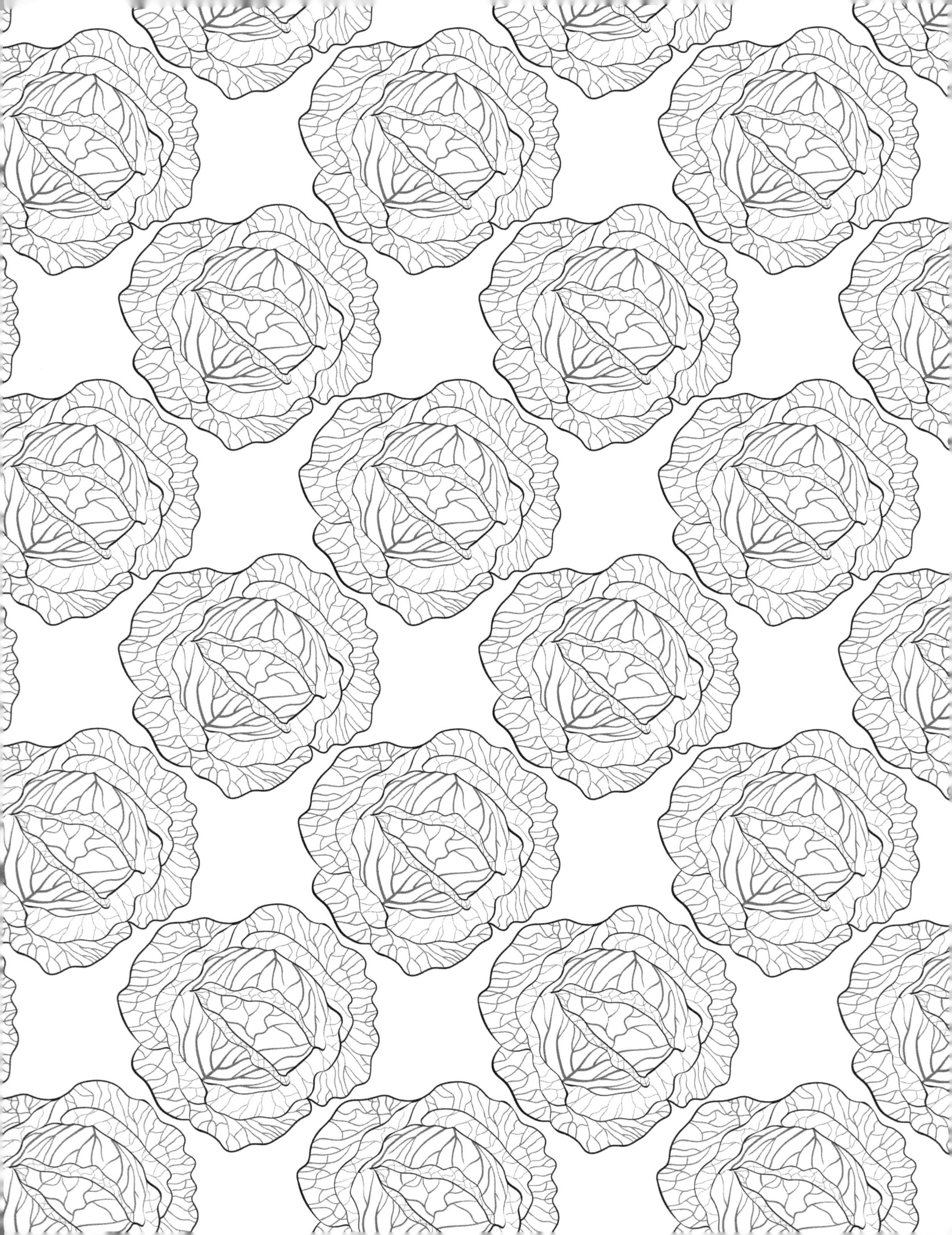

Cabbage

Brassica oleracea

Description

Cabbage is a hardy, green vegetable whose leaves are full of vitamins. However, they are difficult to grow as they prefer cool temperatures and tend to be susceptible to garden pests. Cabbage is prepared many different ways for eating; it can be pickled, fermented (for dishes such as sauerkraut), steamed, stewed, sautéed, braised, or eaten raw. It is a good source of vitamin K, vitamin C and dietary fiber.

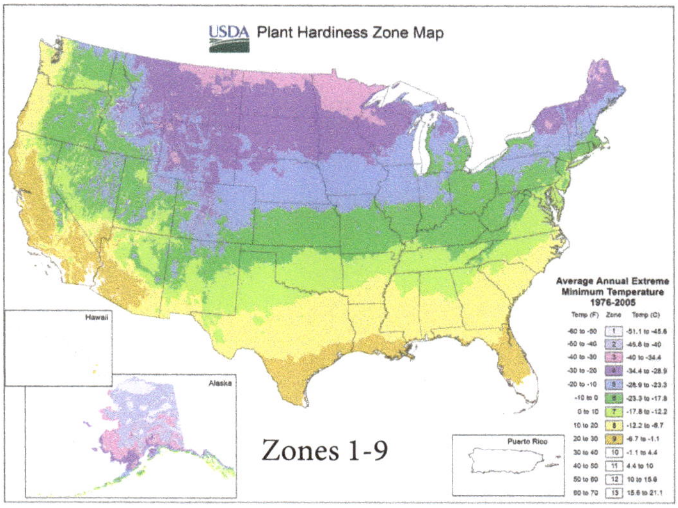

Zones 1-9

Seeding Square Template

1 plant per square foot

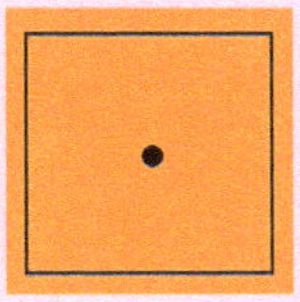

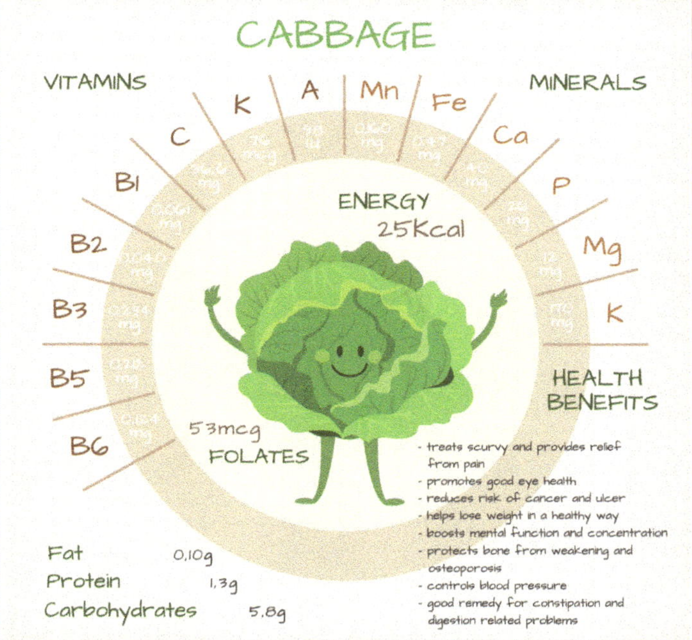

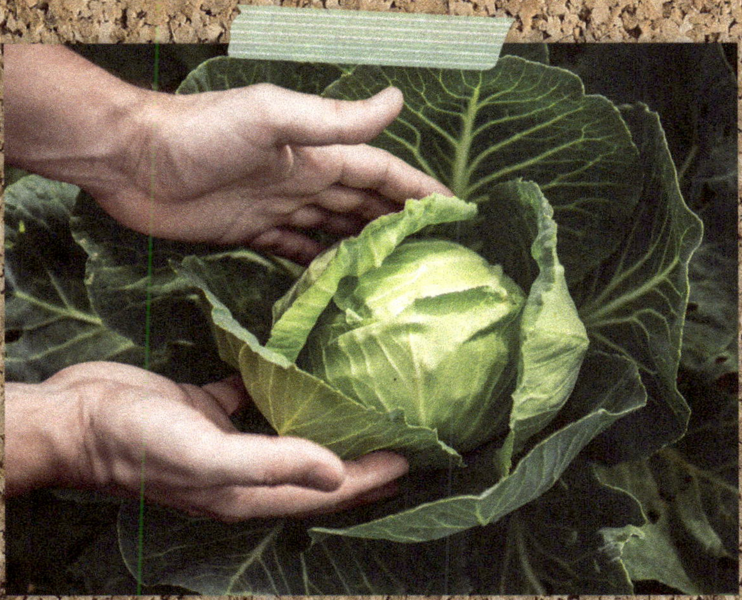

Times to Plant

Start seeds indoors 6-8 weeks before your final frost, then transplant to your garden 2-3 weeks before the final frost.

Companion Plants

Beans, Celery, Cucumbers, Dill, Kale, Lettuce, Onions, Potatoes, Sage, Spinach, Thyme

Cantaloupe

Cucumis melo

Description

Cantaloupe, also known as Muskmelon, is a heat-loving fruit that has a relatively long growing season. Cantaloupe is normally eaten as a fresh fruit, as a salad, or as a dessert with ice cream or custard. Melon pieces wrapped in prosciutto are a familiar antipasto. The seeds are edible and may be dried for use as a snack.

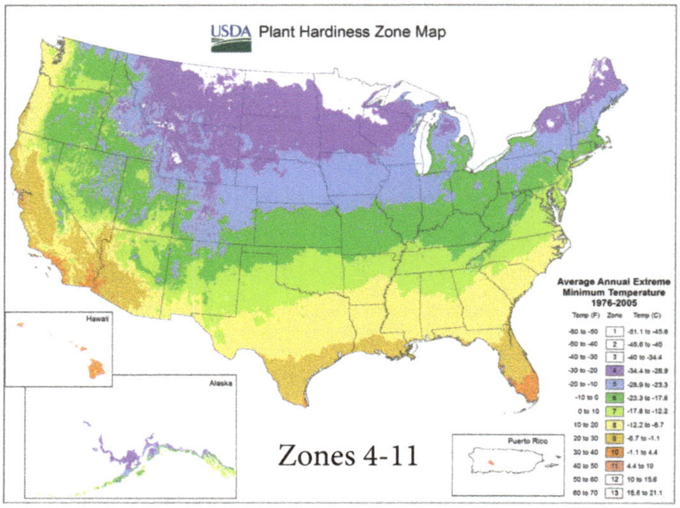

Seeding Square Template

2-4 square feet per melon

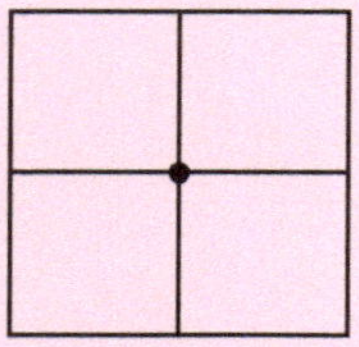

Cantaloupe Nutrition Facts

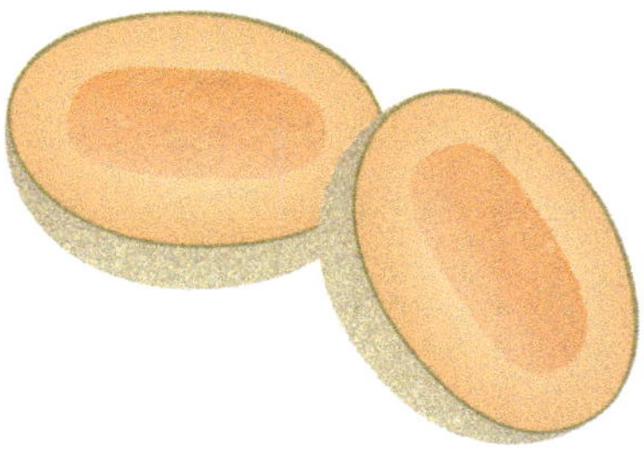

Nutrition Facts	
Serving Size 1 medium	552 g
Amount Per Serving	
Calories 188	Calories from fat 9
	% Daily Value*
Total Fat 1g	2%
Saturated Fat 0g	1%
Trans Fat 0g	
Cholesterol 0g	0%
Sodium 88mg	4%
Total Carbohydrate 49g	16%
Dietary Fiber 5g	20%
Sugars 43g	
Protein 5g	
Vitamin A	373%
Vitamin C	338%
Calcium	5%
Iron	6%

*Percent Daily Values are based on a 2,000 calorie diet. Your daily values may be higher or lower depending on your calorie needs.

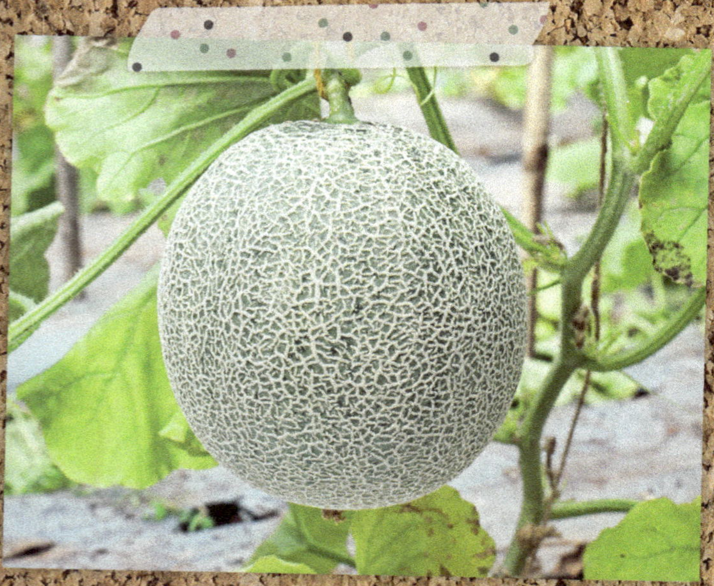

Times to Plant

If you live in a warmer climate, seeds can be sewn directly into your garden once your soil temperatures reach a minimum of 65°.

Companion Plants

Corn, pumpkin, squash, collards, borage, oregano, radishes, marigolds, petunias and beans

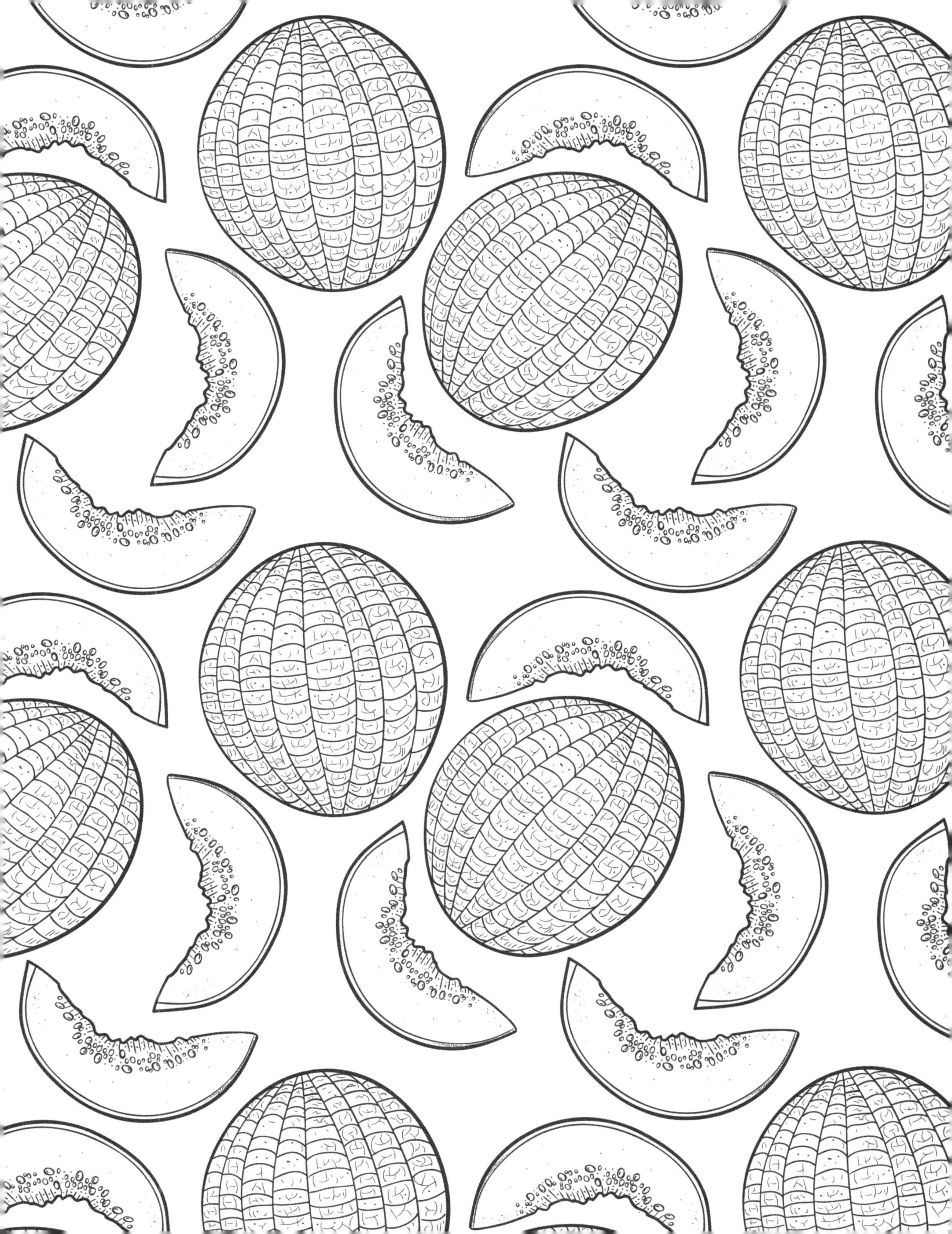

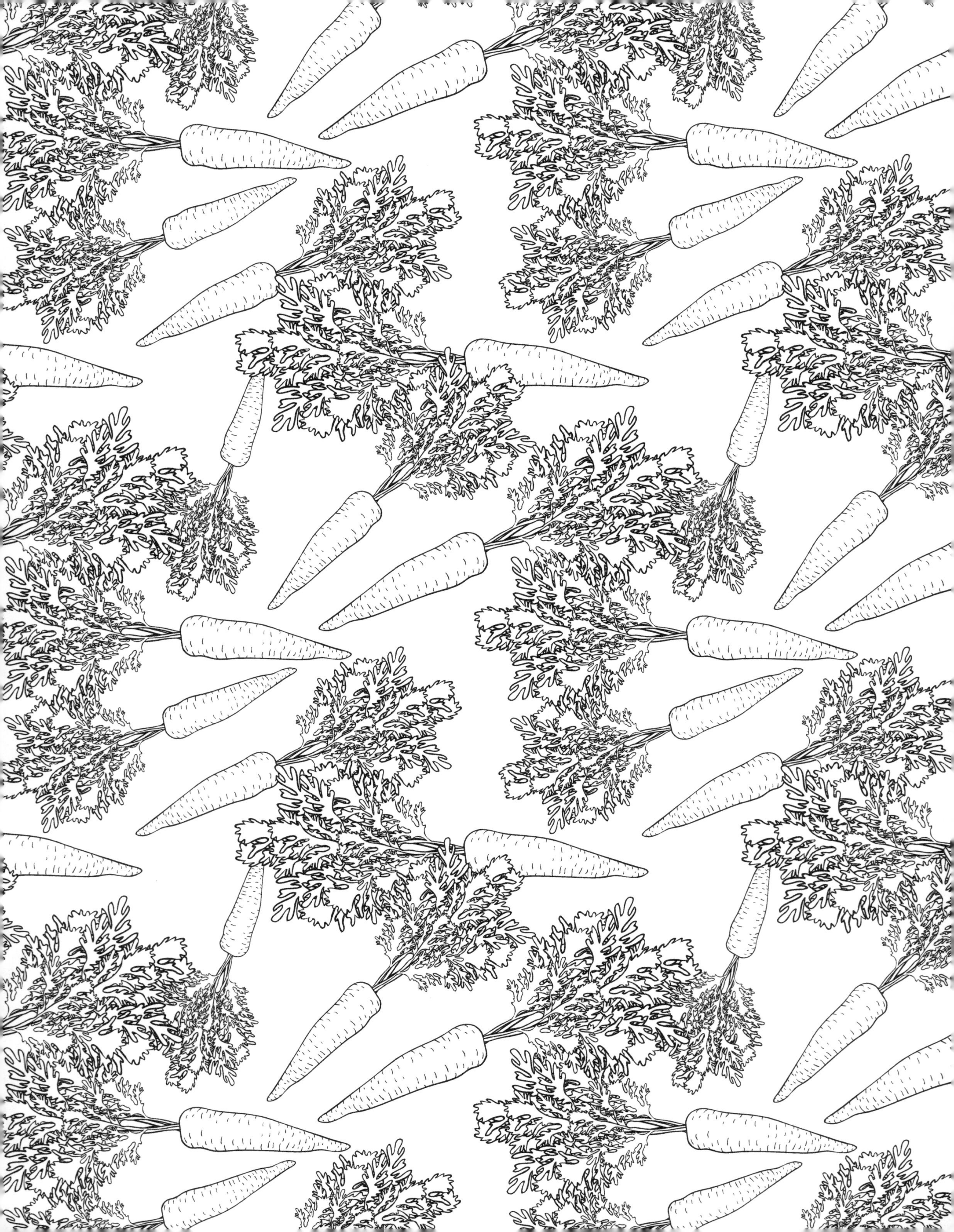

Carrots

Daucus carota

Description

The carrot is a root vegetable, usually orange in color (although purple, black, red, white, and yellow colors exist). Most varieties of carrots are resistant to pests and disease, and are also tolerant of frost. They prefer slightly cooler weather, so they are most often planted in the spring and fall seasons. It is said that the only problem a gardener can encounter with carrots is that their soil could be too heavy. Carrots need light, sandy soil in order to grow properly, otherwise their growth is stunted.

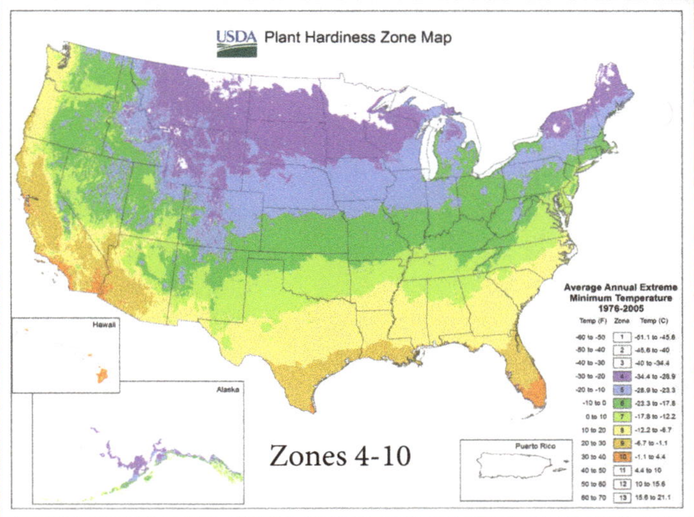

Zones 4-10

Seeding Square Template

Plant seeds every 3", or 16 seeds per square foot.

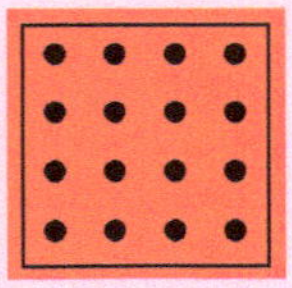

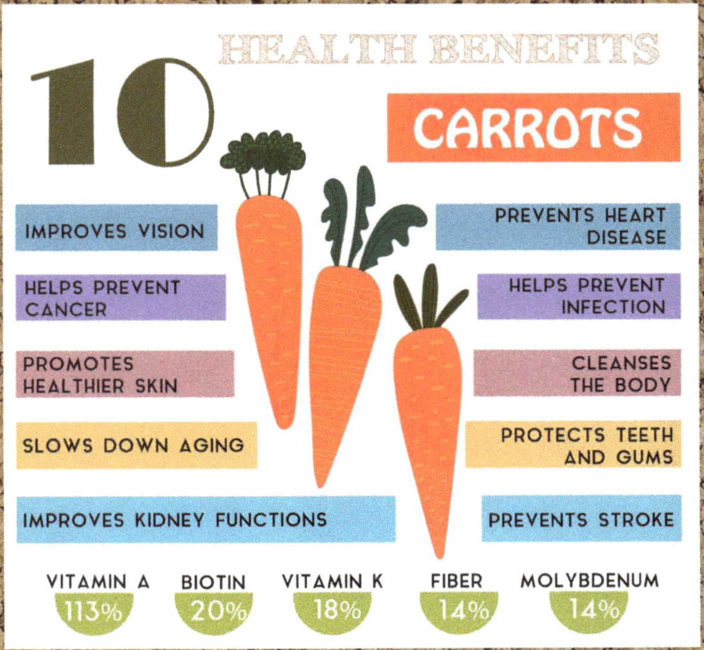

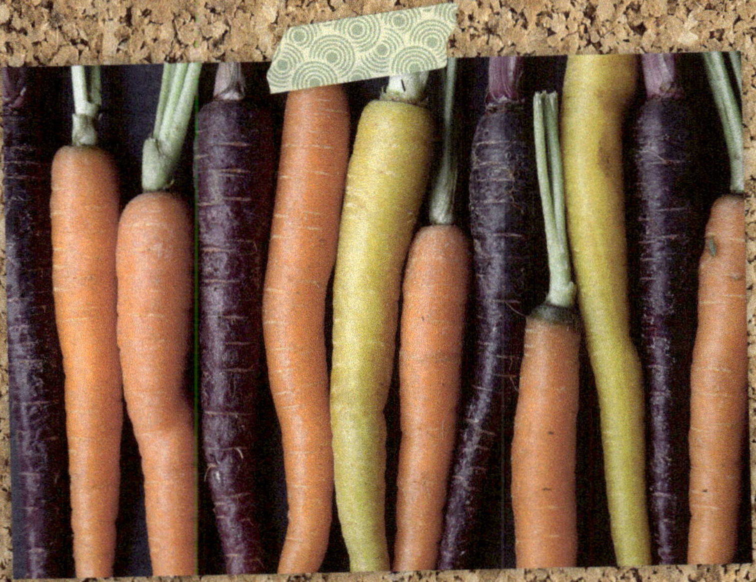

Times to Plant

Plant seeds indoors 3 to 5 weeks before last frost date. They may need up to 4 weeks before they show signs of life. You can also plant them directly outdoors 14 weeks before your first fall frost date.

Companion Plants

Beans, Lettuce, Onions, Peas, Radishes, Rosemary, Sage, Tomatoes, Dill, and Parsley

Cauliflower

Brassica oleracea

Description

Cauliflower is a cool-season crop related to the cabbage; however, it is more difficult to grow than a common cabbage. It requires temperatures consistently around 60º, otherwise it may prematurely bloom, or may produce a poor quality plant. It is a challenging plant, but very rewarding, if able to grow properly.

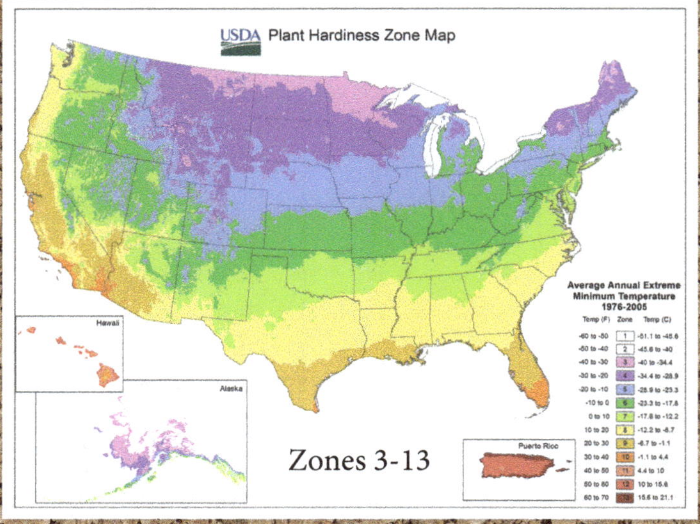

Seeding Square Template

18" per plant, or 4 plants in 9 squares

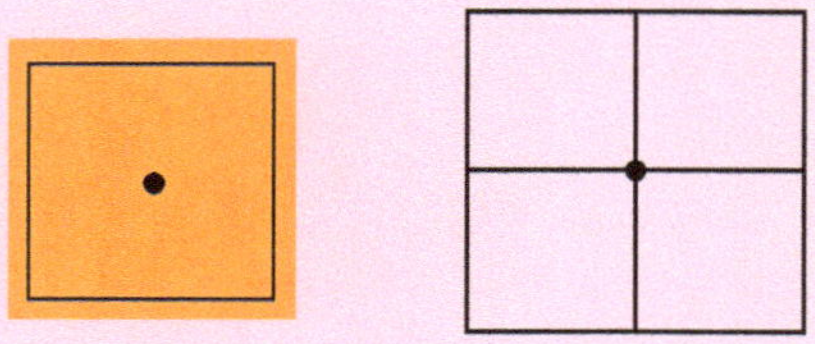

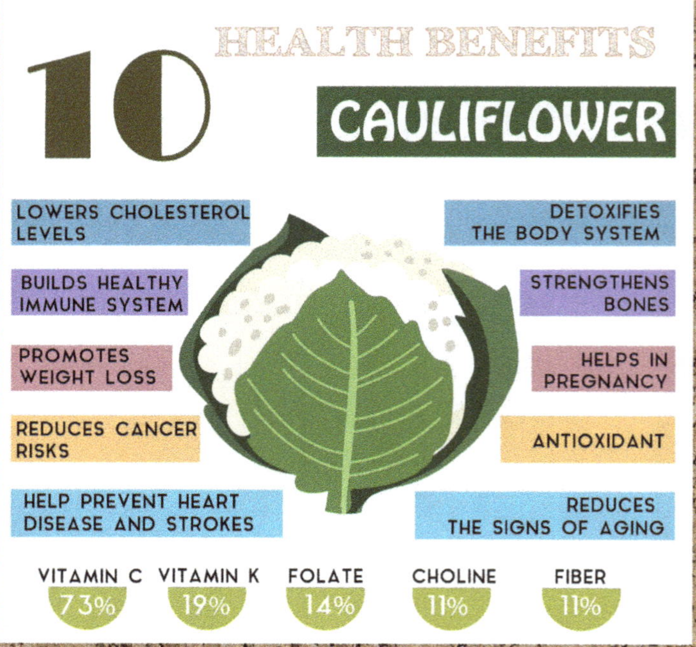

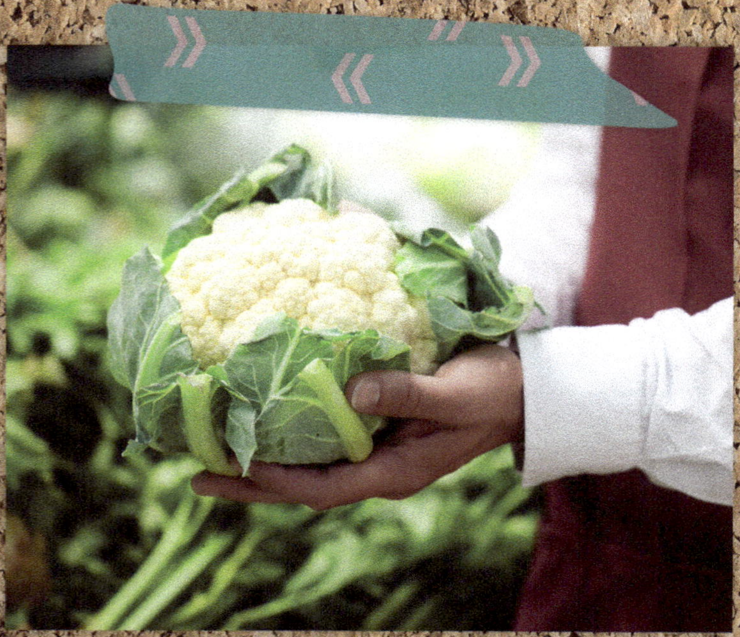

Times to Plant

Start cauliflower as transplants, rather than seeds, 2-4 weeks before your final frost date in the spring.

Companion Plants

Beets, Broccoli, Brussels sprouts, Chard, Spinach, Cucumber, Corn, Radish

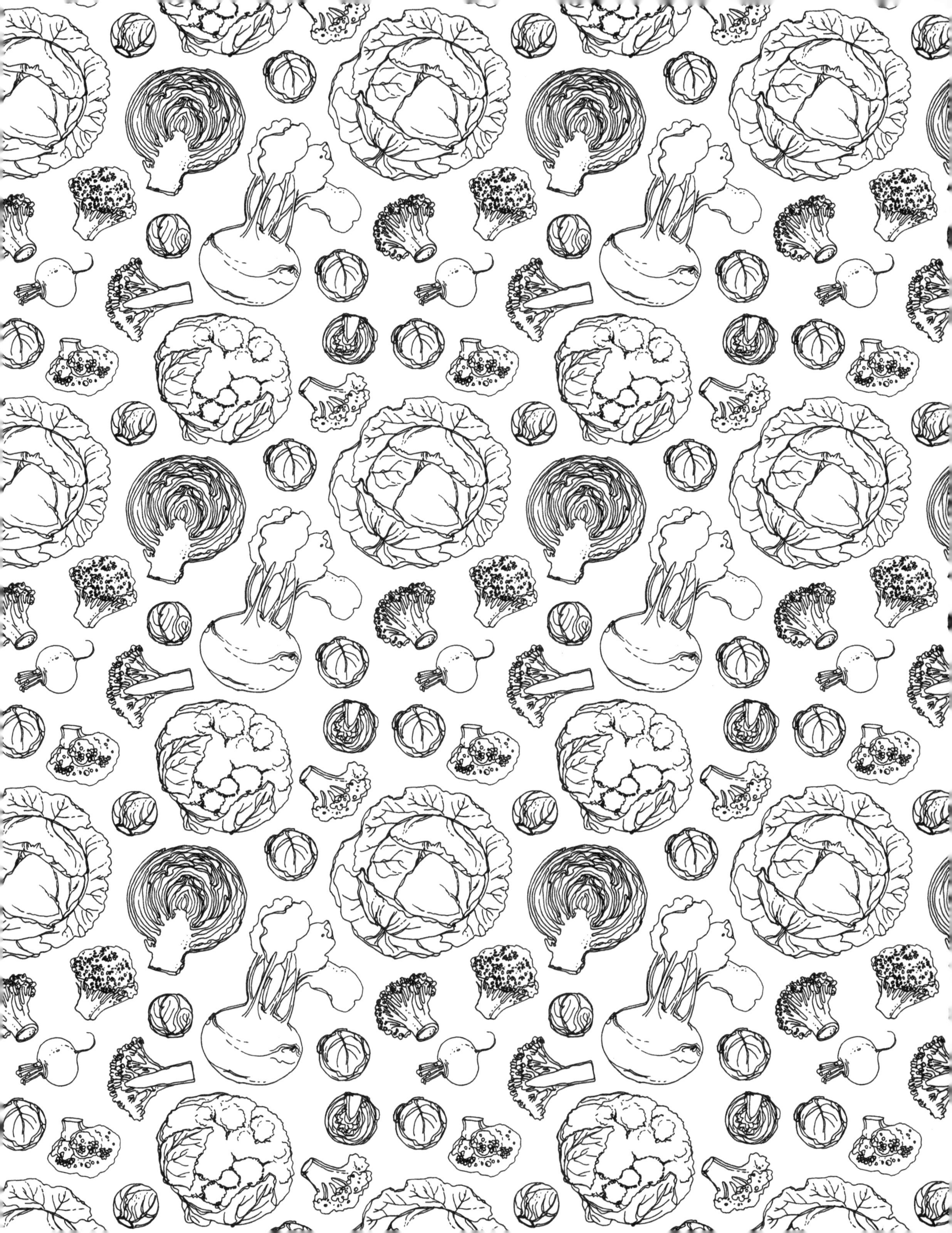

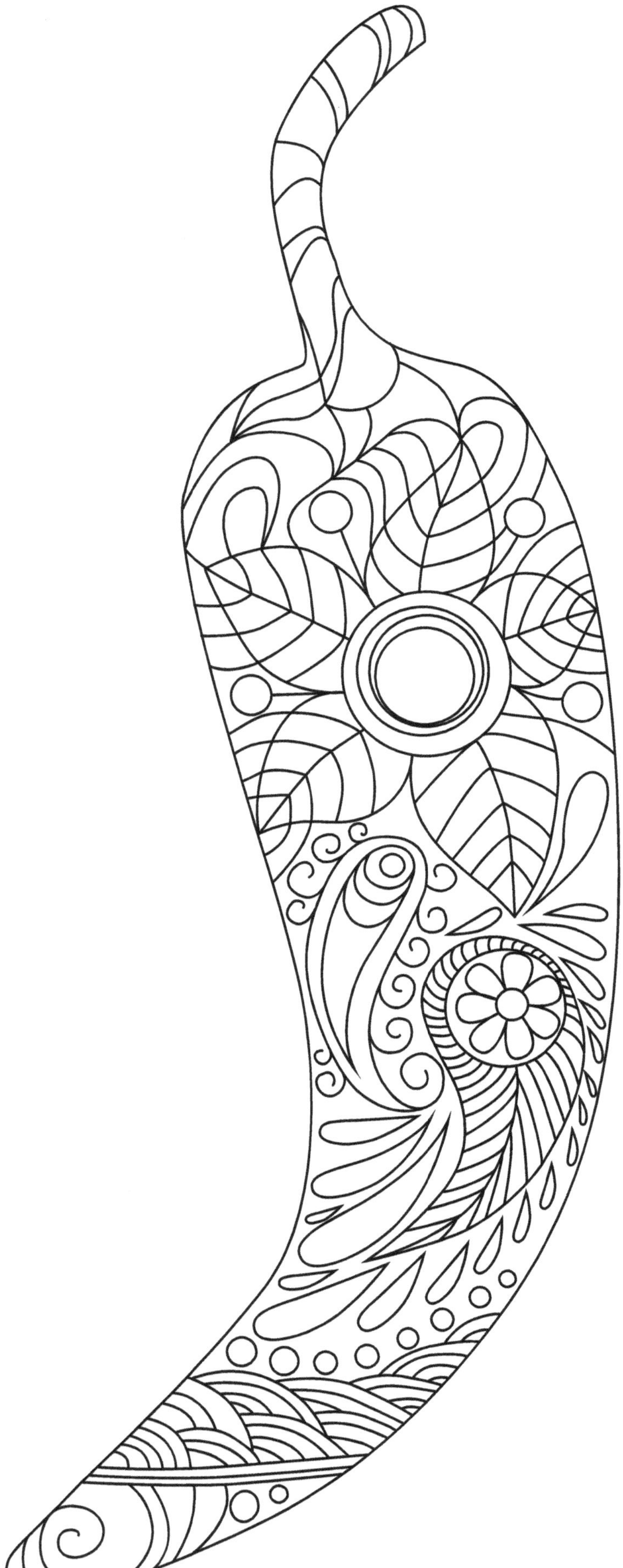

Cayenne

Cayenne Pepper

Capsicum annuum

Description

The cayenne pepper is a moderately hot chili pepper used to flavor dishes, and is also used as an herbal supplement. They are part of the Capsicum family, just like bell and jalapeño peppers. The fruits are generally dried and ground to make the powdered spice of the same name, although cayenne powder may be a blend of different types of peppers, quite often not even containing cayenne peppers. Like bell peppers, cayenne peppers do not tolerate cold weather, and only survive in warm and humid climates.

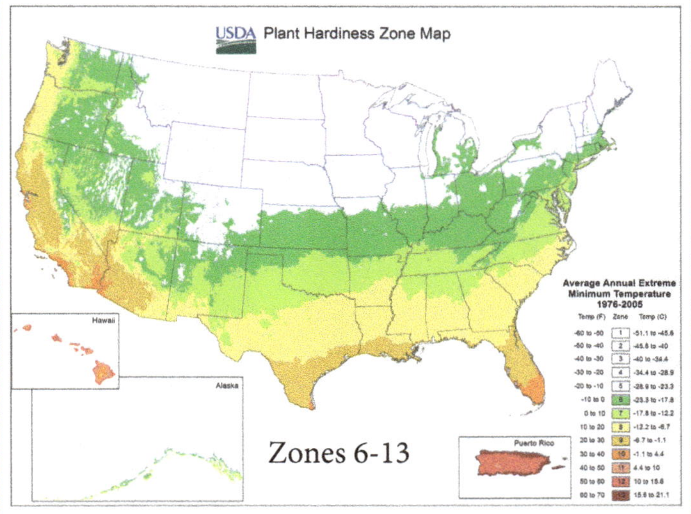

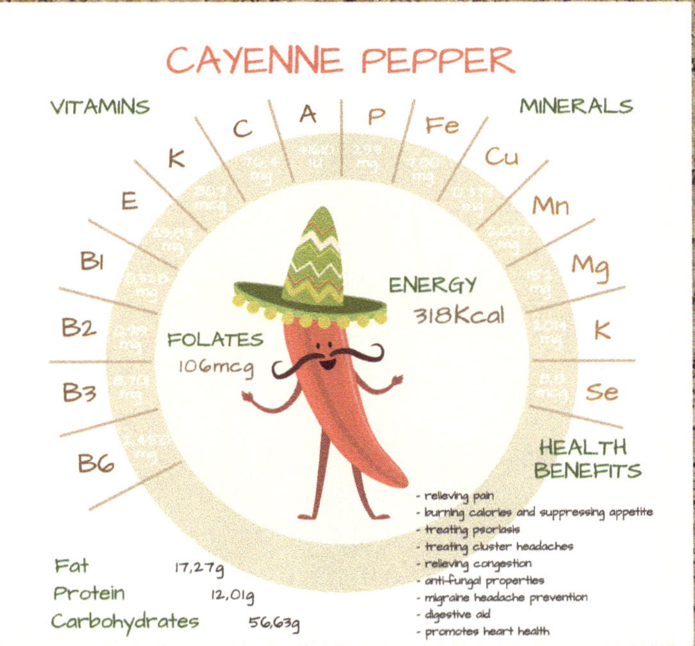

Seeding Square Template

1 plant per square foot

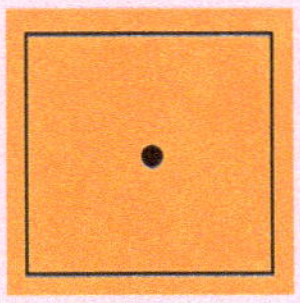

Times to Plant

Plant right before the last frost in your hardiness zone. Using a greenhouse can help you to combat the last frost dates in your zone.

Companion Plants

Cucumbers, eggplant, escarole, tomato, okra, Swiss chard and squash.

Corn

Zea mays

Description

Corn is grown as huge stalks that require lots of sunlight and proper pollination. It has become a staple food in many parts of the world, with the total production of corn surpassing that of wheat or rice. However, little of this corn is consumed directly by humans: most is used for corn ethanol, animal feed, and other maize products, such as corn starch and corn syrup.

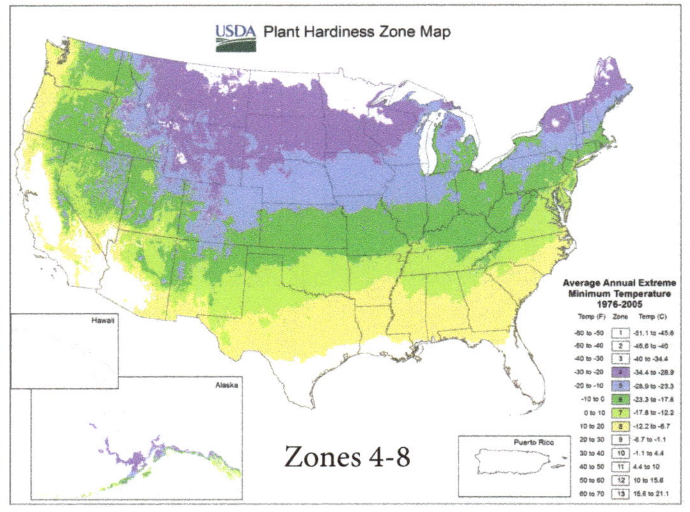

Seeding Square Template

6" per plant, so 4 plants per square foot

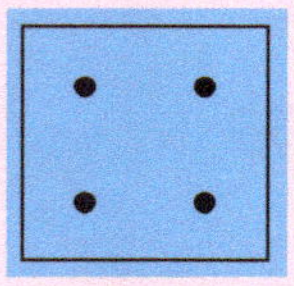

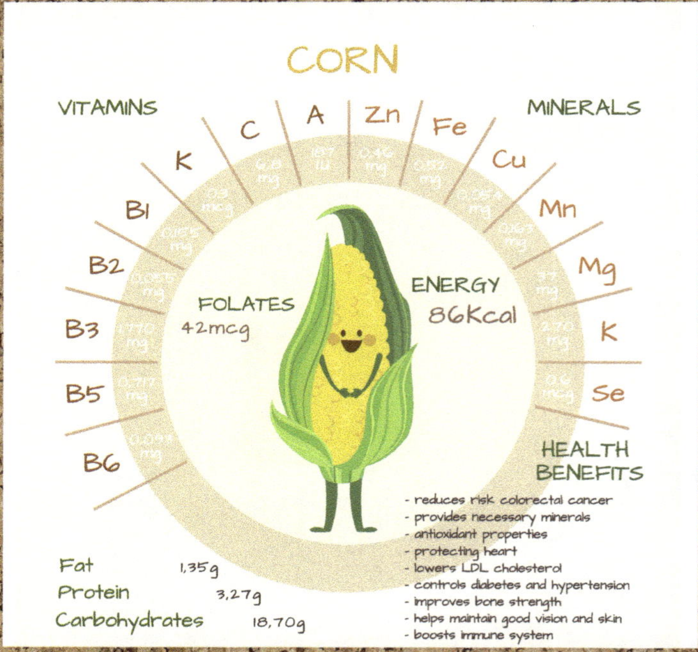

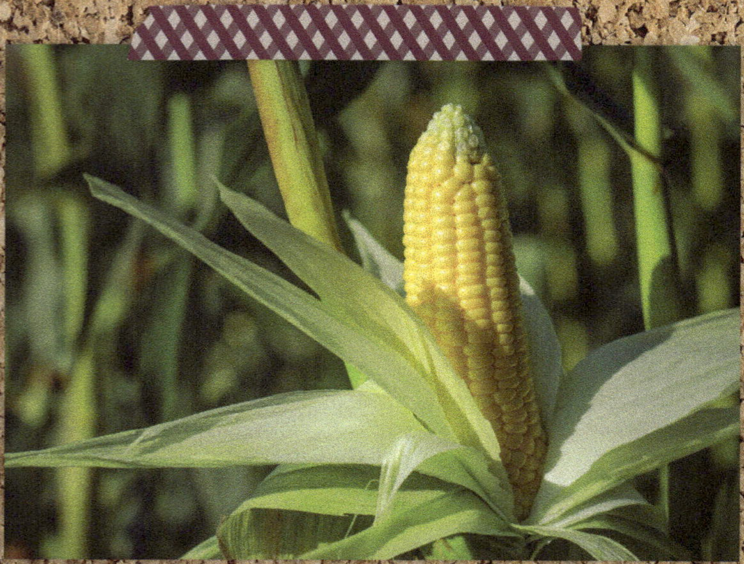

Times to Plant

Plant seeds outdoors two weeks after the last spring frost date, and after the soil is consistently above 60°.

Companion Plants

Beans, Cucumbers, Lettuce, Melons, Peas, Potatoes, Squash, Sunflowers

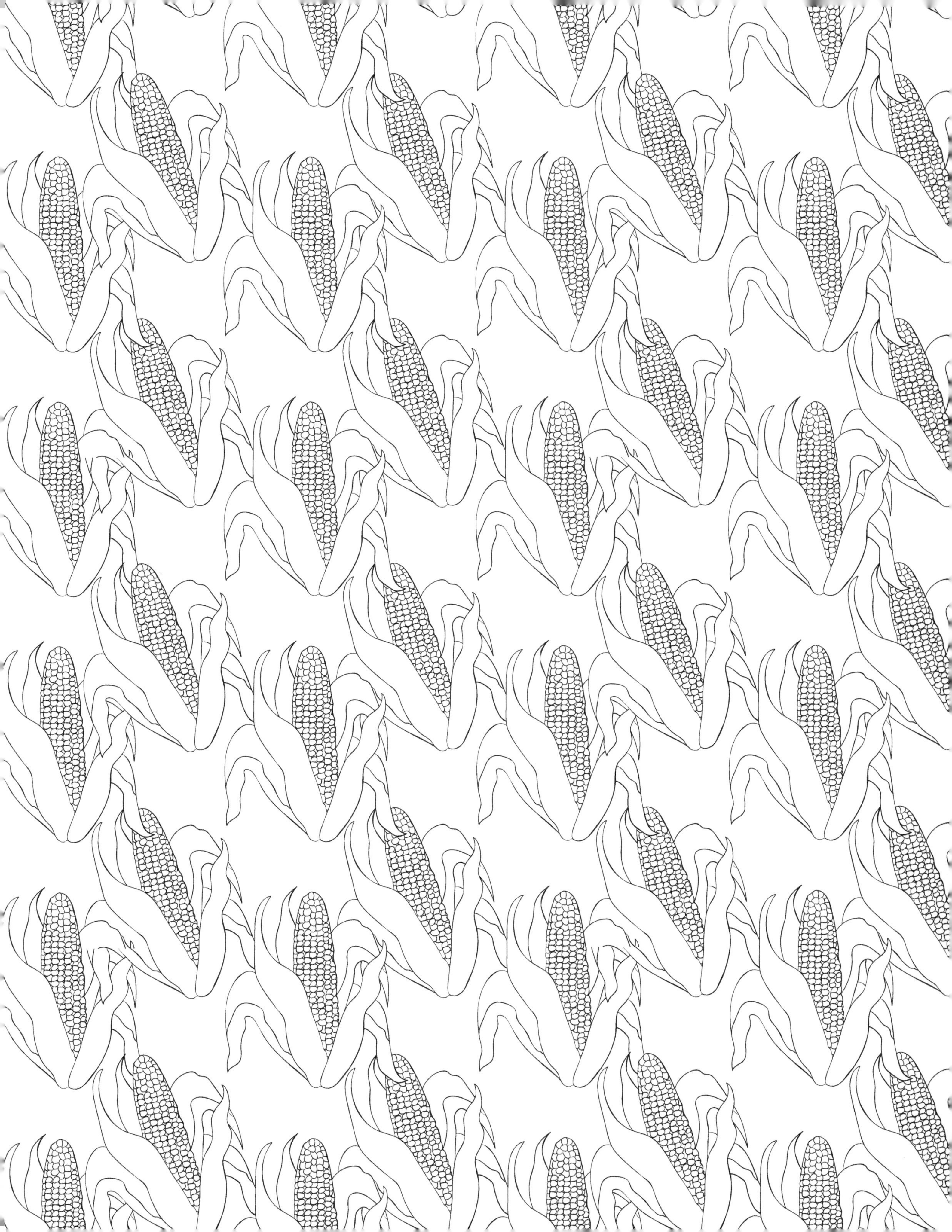

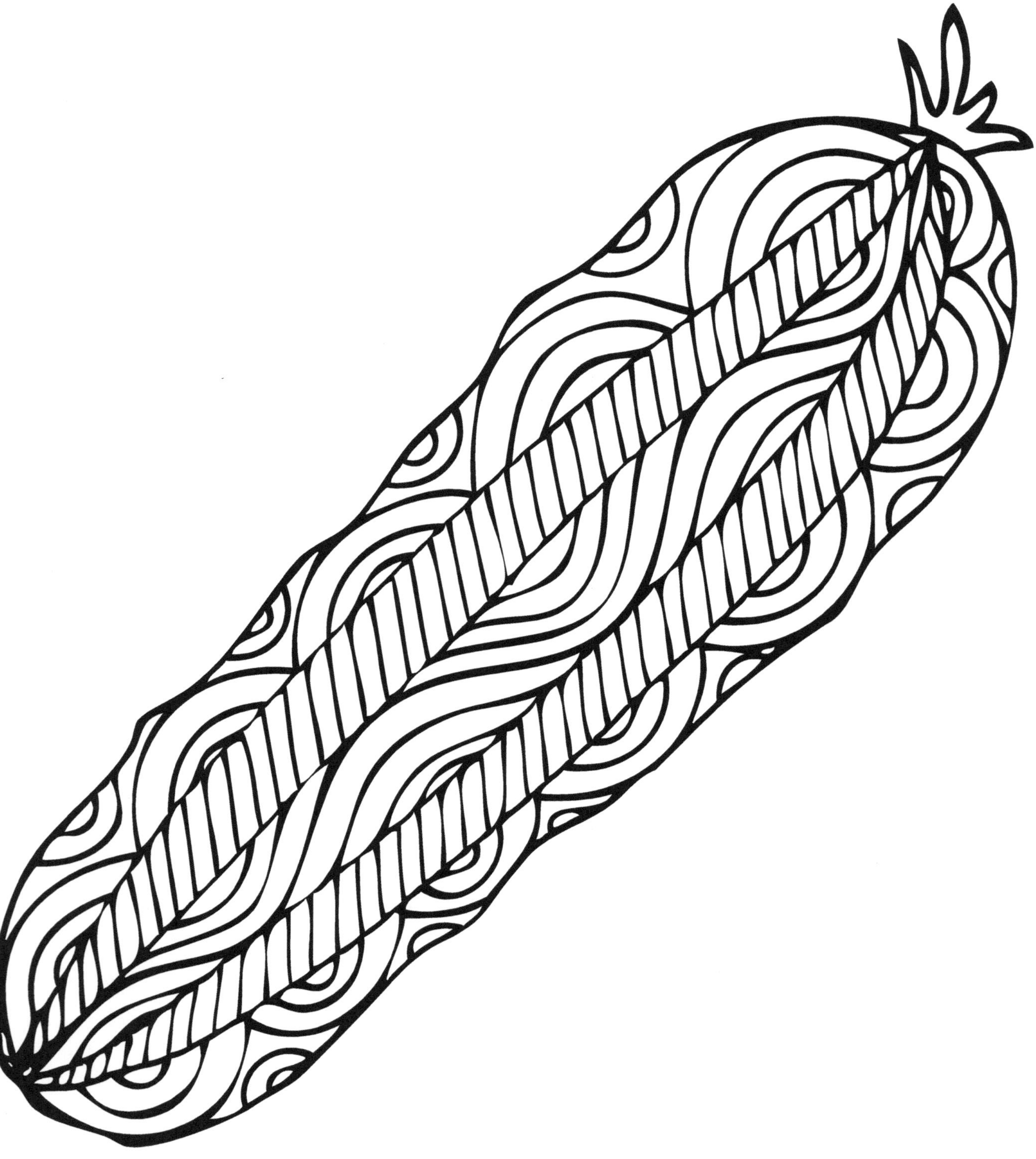

Cucumber

Cucumber

Cucumis sativus

Description

The cucumber is a widely cultivated plant in the gourd family. It is a creeping vine that bears fruits that are used as vegetables. The plants love sun and water, and can grow very quickly as long as they are properly cared for. They require lots of water, sun, and warmth. There are two types of plants: vining and bush. The most commonly grown are the vining cucumbers, which need a trellis or wall to climb, however, bush cucumbers grow well in small containers.

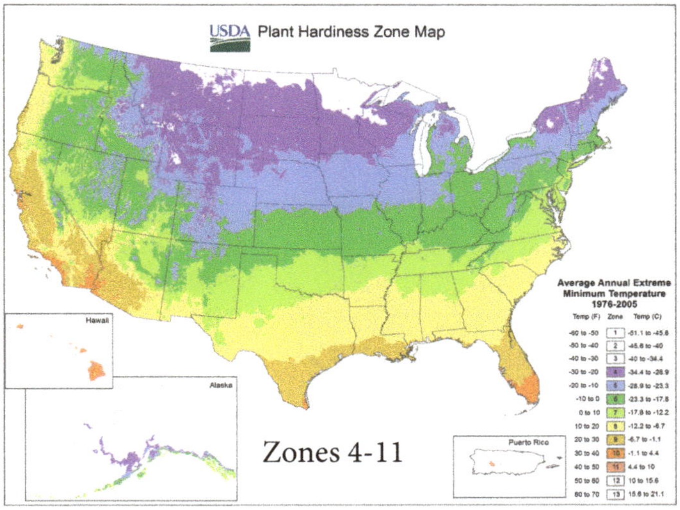

Seeding Square Template

1 plant per 4 square feet

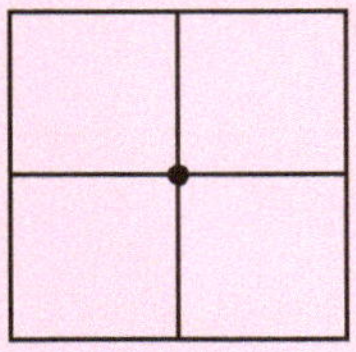

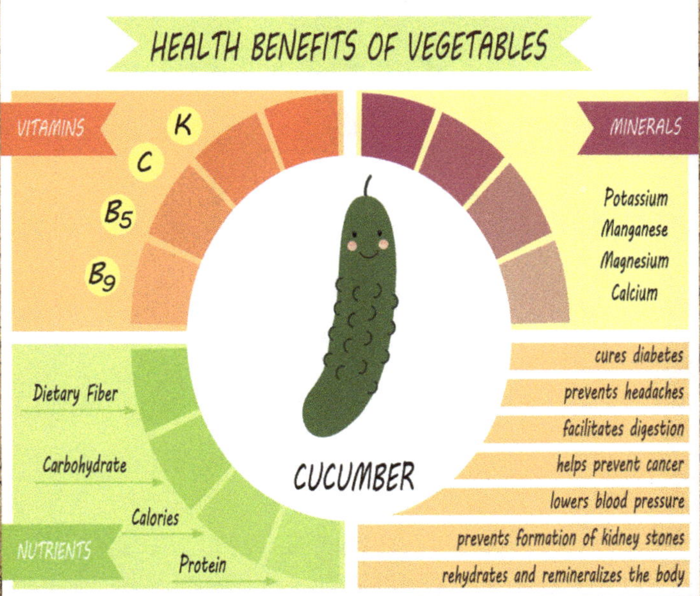

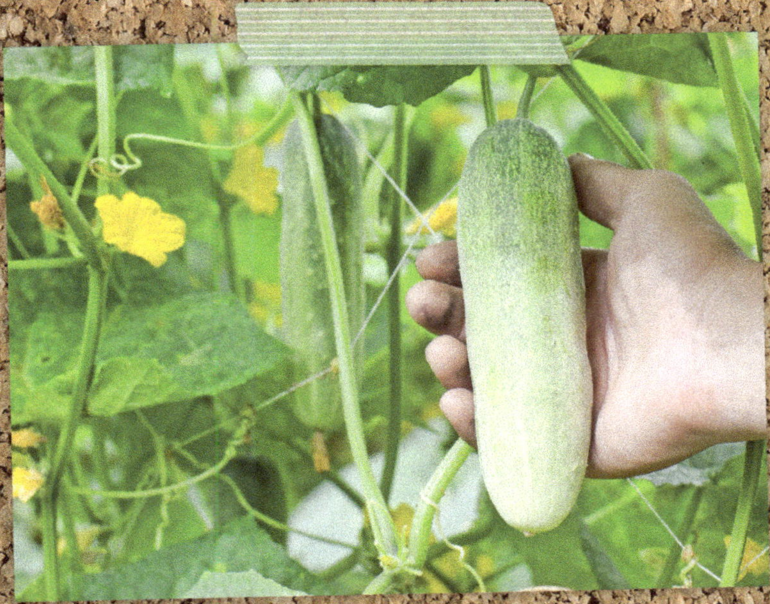

Times to Plant

Cucumbers are extremely susceptible to cold temperatures, so it is important to plant them (at the earliest) two weeks after your final frost date. They cannot be grown in the fall and winter months.

Companion Plants

Beans, Broccoli, Cabbage, Cauliflower, Corn, Lettuce, Peas, Radishes, Sunflowers, Herbs, Melons, and Potatoes

Eggplant

Solanum melongena

Description

The Eggplant is a species in the nightshade family. It is grown for its edible spongy absorbent fruit widely used in cooking. Most people consider it a vegetable despite its botanical definition as a berry fruit. Like the tomato, its skin and seeds can be eaten, but like the potato, it is not to be eaten raw.

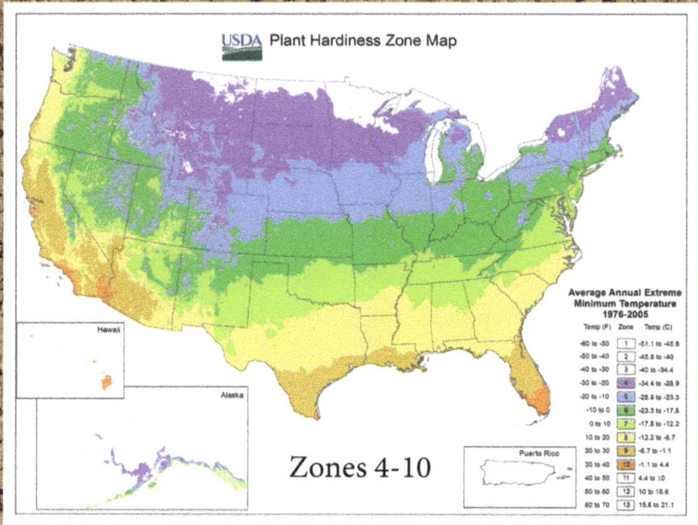

Seeding Square Template

1 plant per square foot

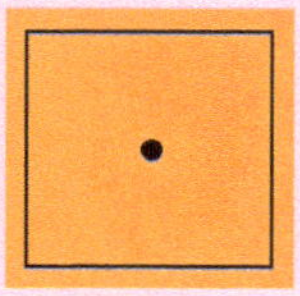

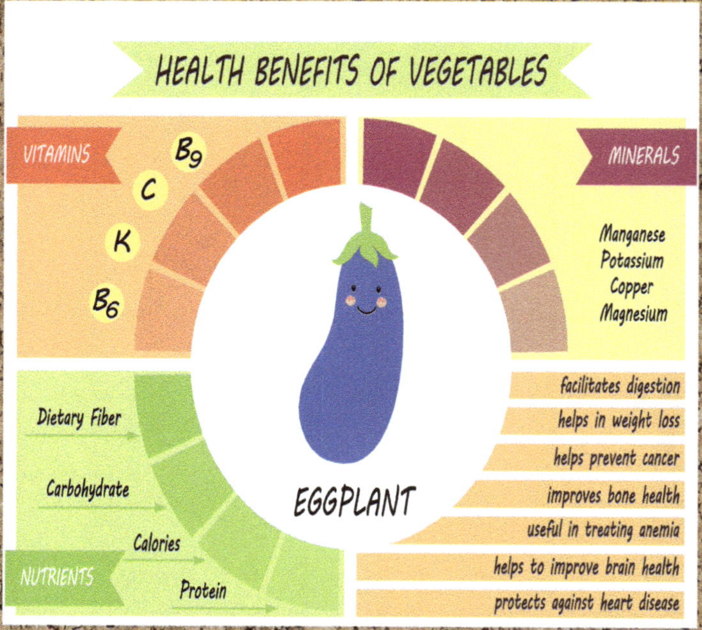

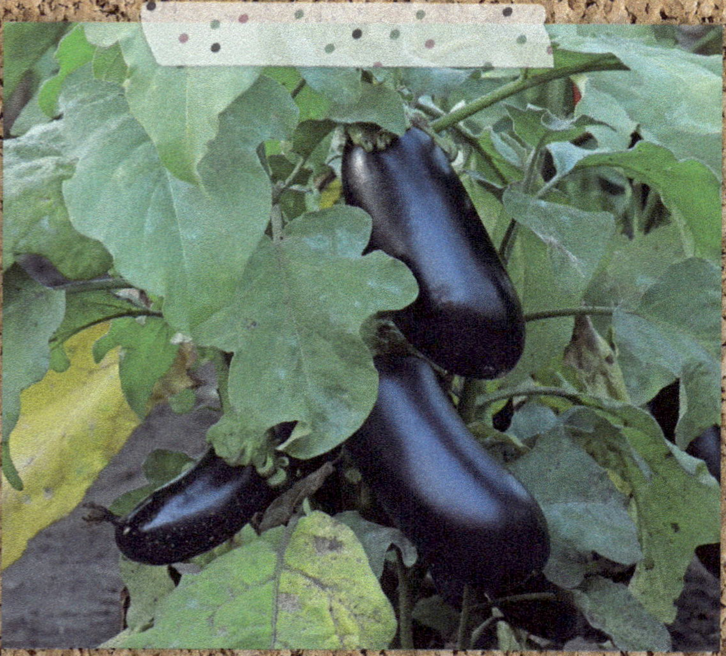

Times to Plant

Start seeds indoors 7 to 8 weeks before the frost date, and transplant them into the garden 2 to 4 weeks after the spring frost date.

Companion Plants

Basil, Beans, Lettuce, Peas, Potatoes, Spinach

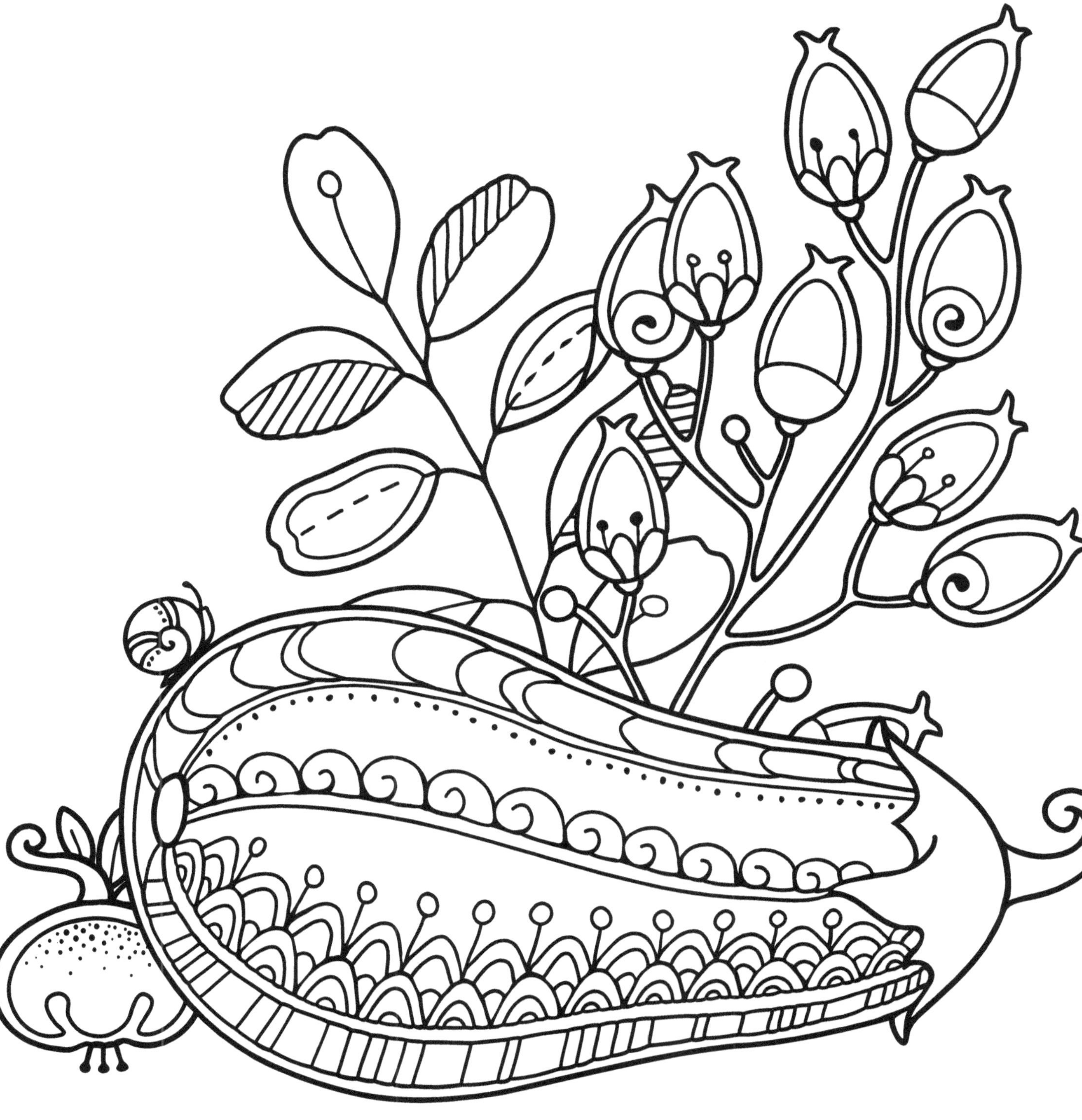

Eggplant

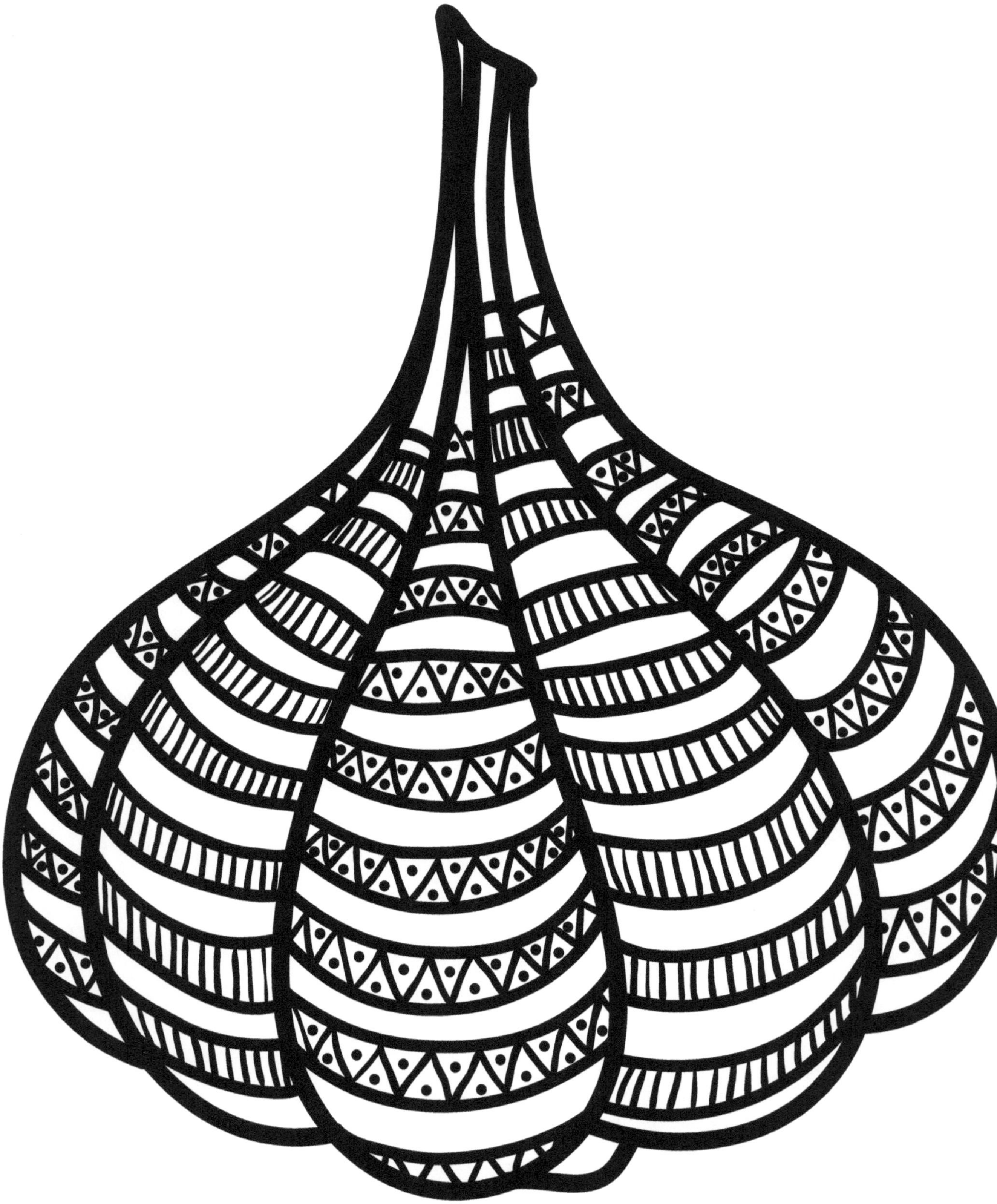

Garlic

Garlic

Allium sativum

Description

Garlic is a bulb that contains individual cloves. These cloves can be used for cooking, planting more garlic, or holistic medicine. They are planted in the fall time and are kept underground until spring, where they grow greenery and their bulbs grow in size.

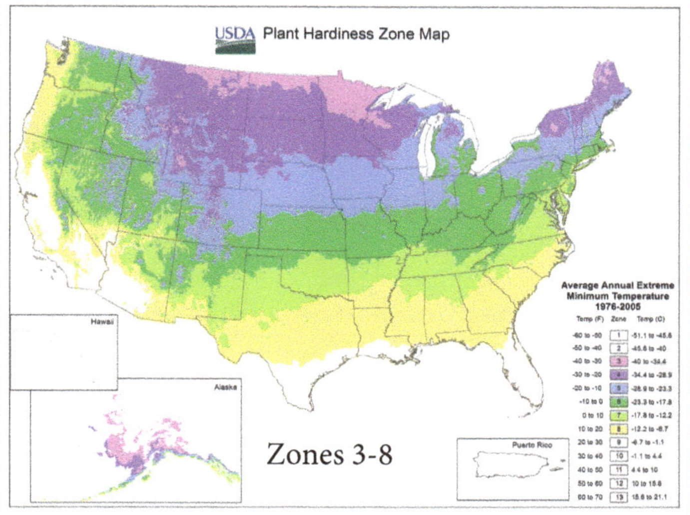

Seeding Square Template

4 cloves per square foot

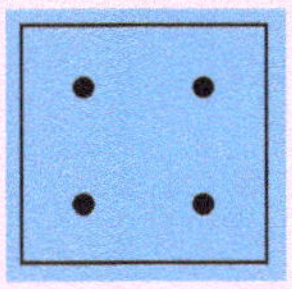

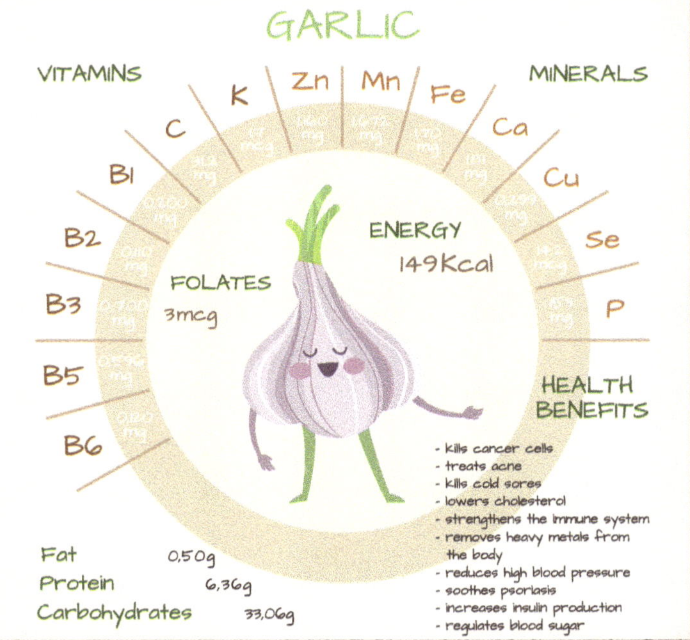

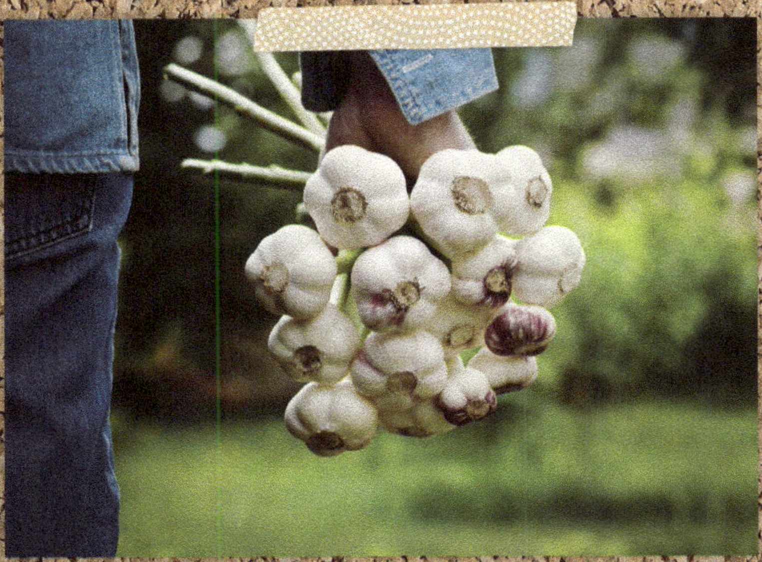

Times to Plant

In areas that get a hard frost, plant 6-8 weeks before your first Fall frost. Your goal is to give the plant long enough to develop a root system before the freeze, but not enough time to develop top growth.

Companion Plants

Beets, strawberry, tomato, lettuce, summer savory, chamomile, roses

Honeydew melon

Cucumis melo

Description

Unique for their smooth, cream-colored skin, Honeydew melons take a bit longer than other melons to grow. Their growth time is, on average, 120 days after germination. Like most fruit, honeydew has seeds. The inner flesh is eaten, often for dessert, and honeydew is commonly found in supermarkets across the world alongside cantaloupe melons and watermelons.

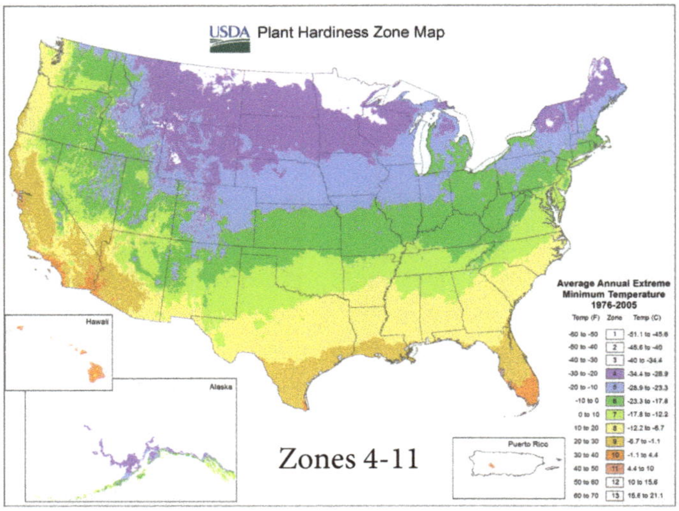

Zones 4-11

Seeding Square Template

2-4 square feet per seed

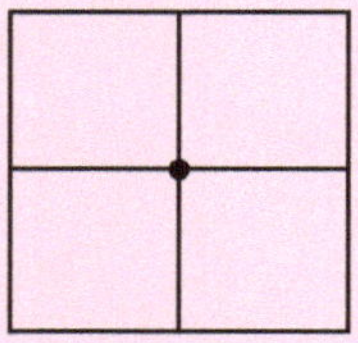

Honeydew Melon

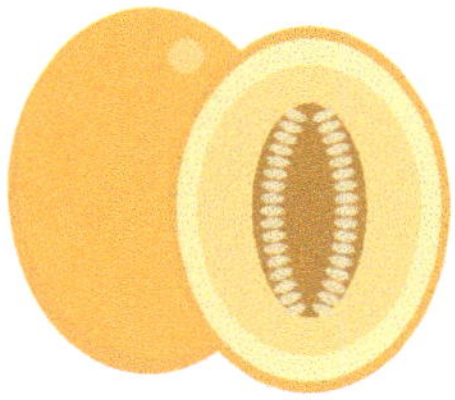

Nutrition Facts

Serving Size 1/2 Cup

Amount Per Serving	
Calories 32	Calories from Fat 1
	% Daily Values*
Total Fat 0g	**0%**
Saturated Fat 0g	**0%**
Trans Fat 0g	
Cholesterol 0mg	**0%**
Sodium 16mg	**1%**
Total Carbohydrate 8g	**3%**
Dietary Fiber 1g	**4%**
Sugars 7g	
Protein 0g	**0%**
Vitamin A 1% •	Vitamin C 27%
Calcium 1% •	Iron 1%

*Percent Daily Values are based on a 2,000 calorie diet. Your Daily Values may be higher or lower depending on your calorie needs.

	Calories	2,000	2,500
Total Fat	Less than	65g	80g
Sat Fat	Less than	20g	25g
Cholesterol	Less than	300mg	300mg
Sodium	Less than	2400mg	2400mg
Total Carbohydrate		300g	375g
Dietary Fiber		25g	30g

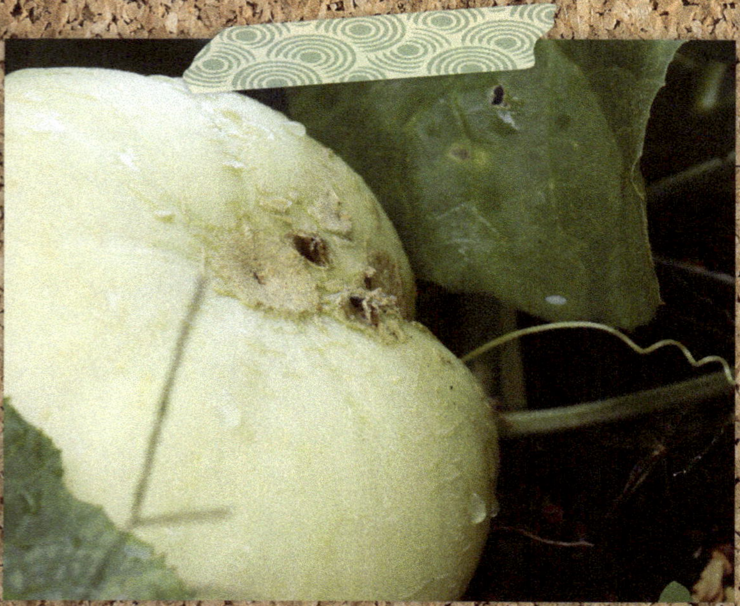

Times to Plant

Seeds can be planted directly in your garden 2-3 weeks after your final frost date, when your average soil temperatures are 70º and above.

Companion Plants

Sunflower, Radishes, Beets, Corn

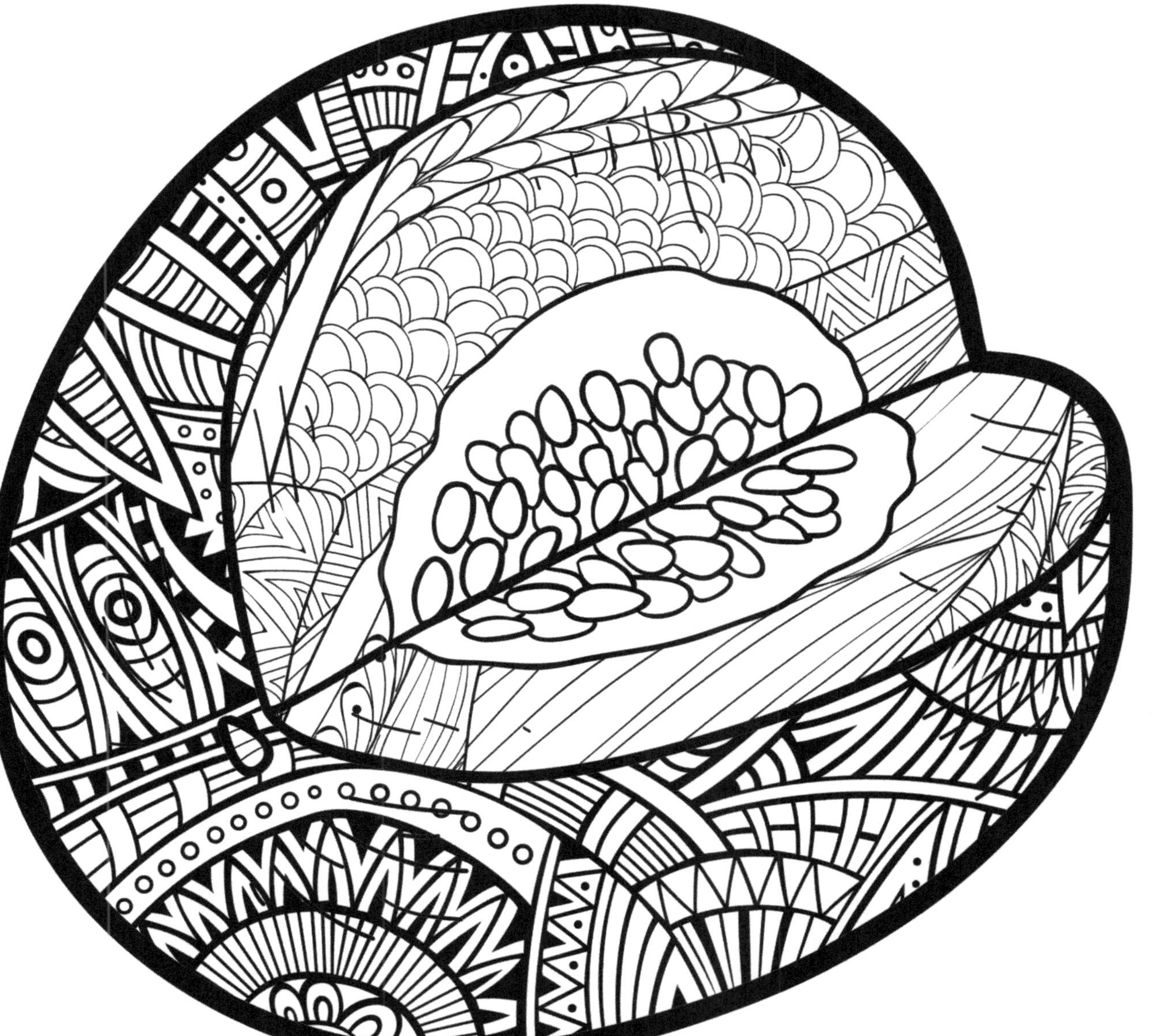

Honeydew Melon

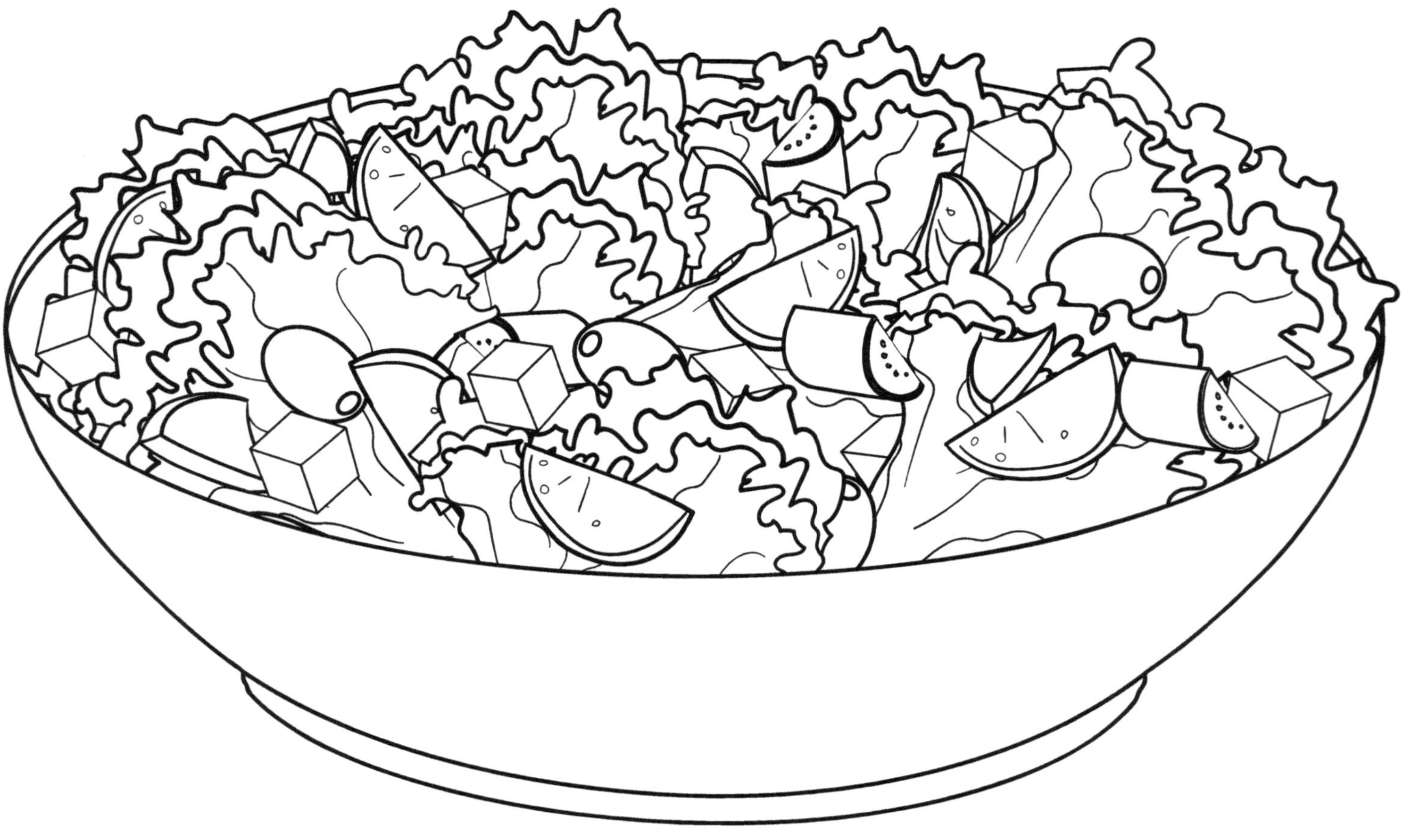

Lettuce

Lettuce

Lactuca sativa

Description

Lettuce is an annual plant of the daisy family, Asteraceae. It is most often grown as a leaf vegetable, but sometimes for its stem and seeds. Growing your own lettuce increases not only the quality and taste, but also the vitamin A content of your plants. Like onions, lettuce is a cold-season crop that prefers the spring and fall seasons. It grows quickly, so most gardeners stagger their plants, and plant a few more seeds every few weeks.

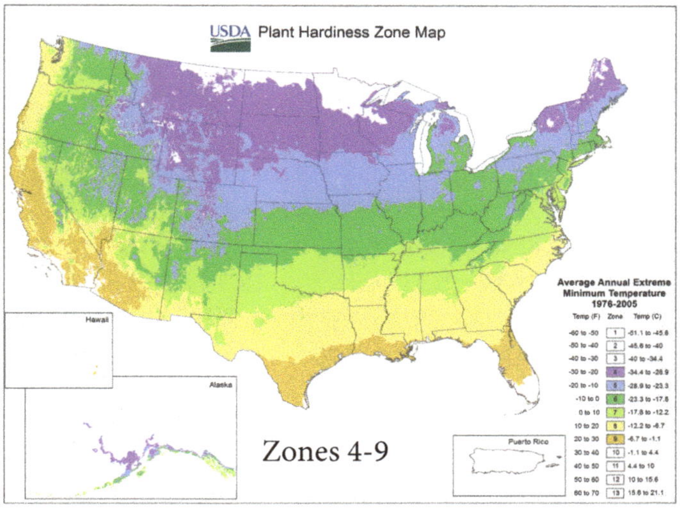

Zones 4-9

Seeding Square Template

6" per seed, so 4 seeds per square foot.

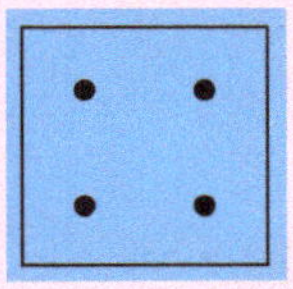

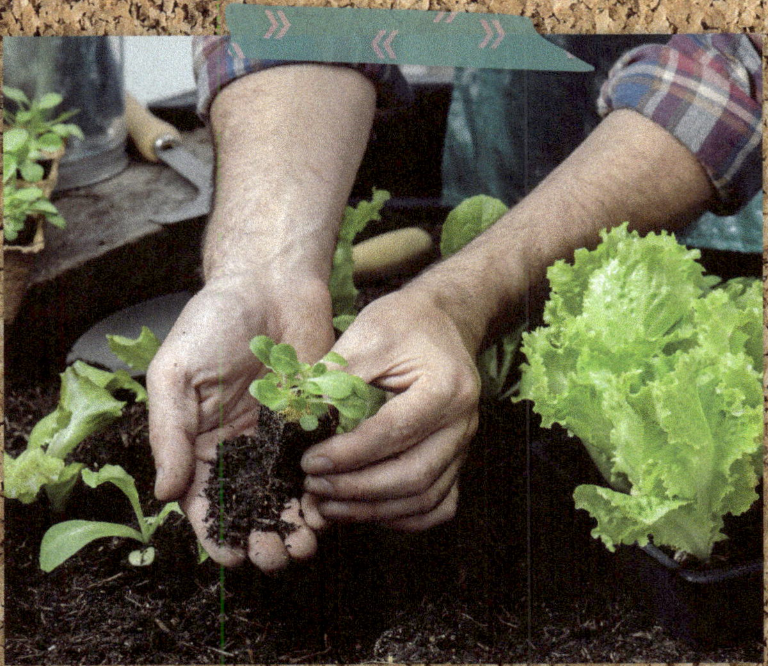

Lettuce

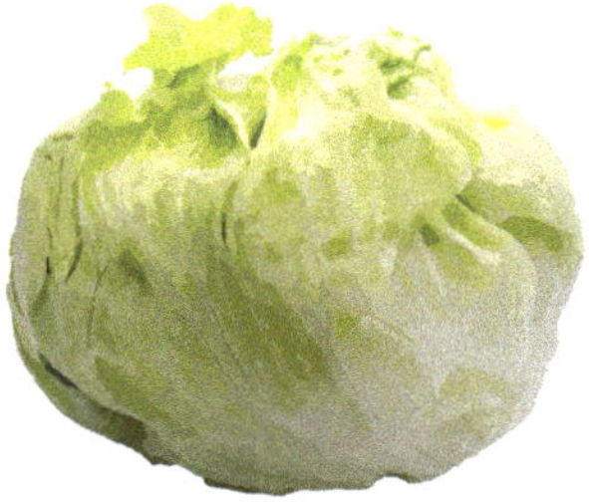

Nutrition Facts

Serving Size 1 large head 755 g

Amount Per Serving	
Calories 106	Calories from fat 9
	% Daily Value*
Total Fat 1g	2%
Saturated Fat 10g	1%
Trans Fat 0g	
Cholesterol 0g	0%
Sodium 75mg	3%
Total Carbohydrate 24g	8%
Dietary Fiber 9g	36%
Sugars 15g	
Protein 7g	
Vitamin A	76%
Vitamin C	35%
Calcium	14%
Iron	17%

*Percent Daily Values are based on a 2,000 calorie diet. Your daily values may be higher or lower depending on your calorie needs.

Times to Plant

Start seedlings indoors 2 or 3 weeks before your final frost date, then transport to your garden once your soil is warm enough to be worked.

Companion Plants

Carrot, cucumber, onion, radish, strawberries, beetroot, marigolds

Okra

Abelmoschus esculentus

Description

Okra is a flowering plant in the mallow family, valued for its edible green seed pods, similar to green bean pods. It is related to such species as cotton, cocoa, and hibiscus. The products of the Okra plant are mucilaginous, resulting in a characteristic "goo" or slime when the seed pods are cooked. Pods are cooked, pickled, eaten raw, or included in salads.

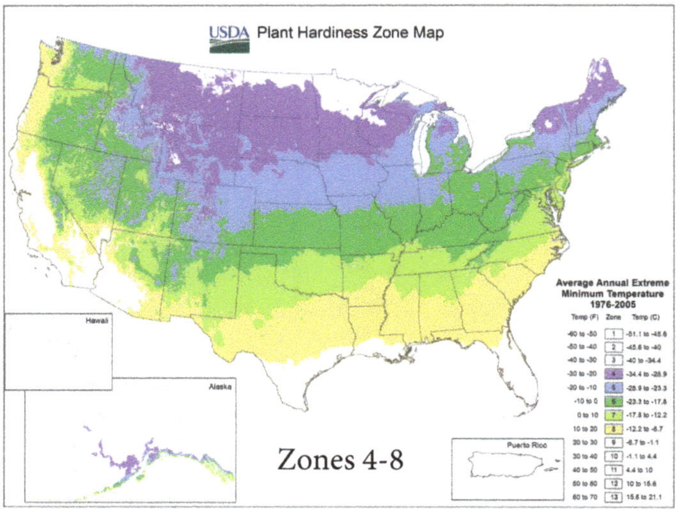

Seeding Square Template

1 plant per square foot

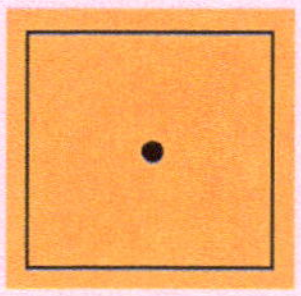

Okra

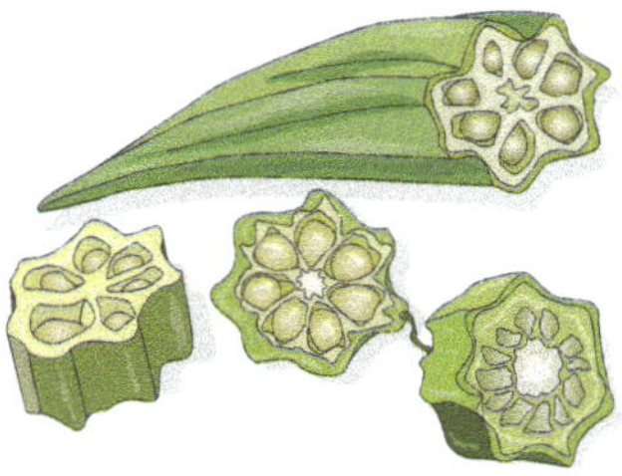

Nutrition Facts

Serving Size 100 grams (100 grams)

Amount Per Serving	
Calories 30	Calories from Fat 2
	% Daily Value*
Total Fat 0g	0%
Saturated Fat 0g	0%
Trans Fat 0g	
Cholesterol 0mg	0%
Sodium 3mg	0%
Total Carbohydrates 7g	2%
Dietary Fiber 2g	9%
Sugars 3g	
Protein 2g	

Vitamin A	7%	Vitamin C	21%
Calcium	8%	Iron	3%

*Percent Daily Values are based on a 2,000 calorie diet. Your daily values may be higher or lower depending on your calorie needs:

	Calories	2,000	2,500
Total Fat	Less than	65g	80g
Sat Fat	Less than	20g	25g
Cholesterol	Less than	300mg	300mg
Sodium	Less than	2,400mg	2,400mg
Total Carbohydrate		300g	375g
Fiber		25g	30g

Calories per gram:
Fat 9 • Carbohydrate 4 • Protein 4

Ingredients: Whole, Okra Frozen, Unprepared

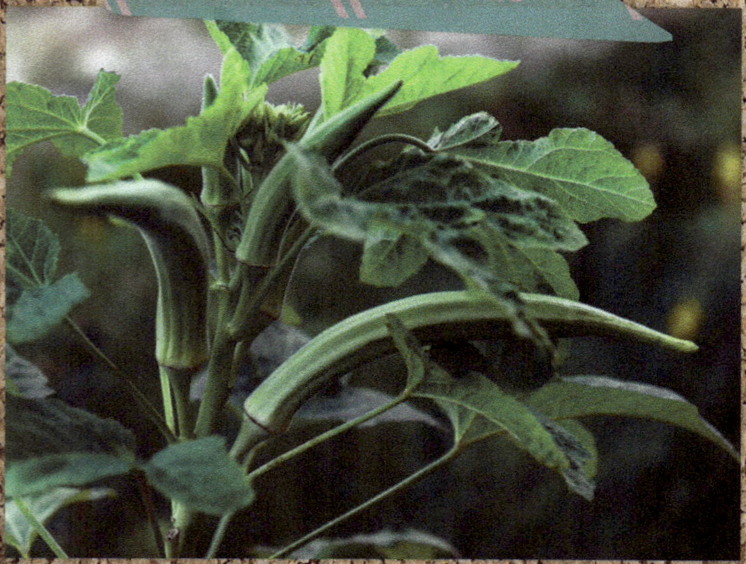

Times to Plant

3-4 weeks before the final frost date

Companion Plants

Melons, cucumbers, sweet peppers, eggplant

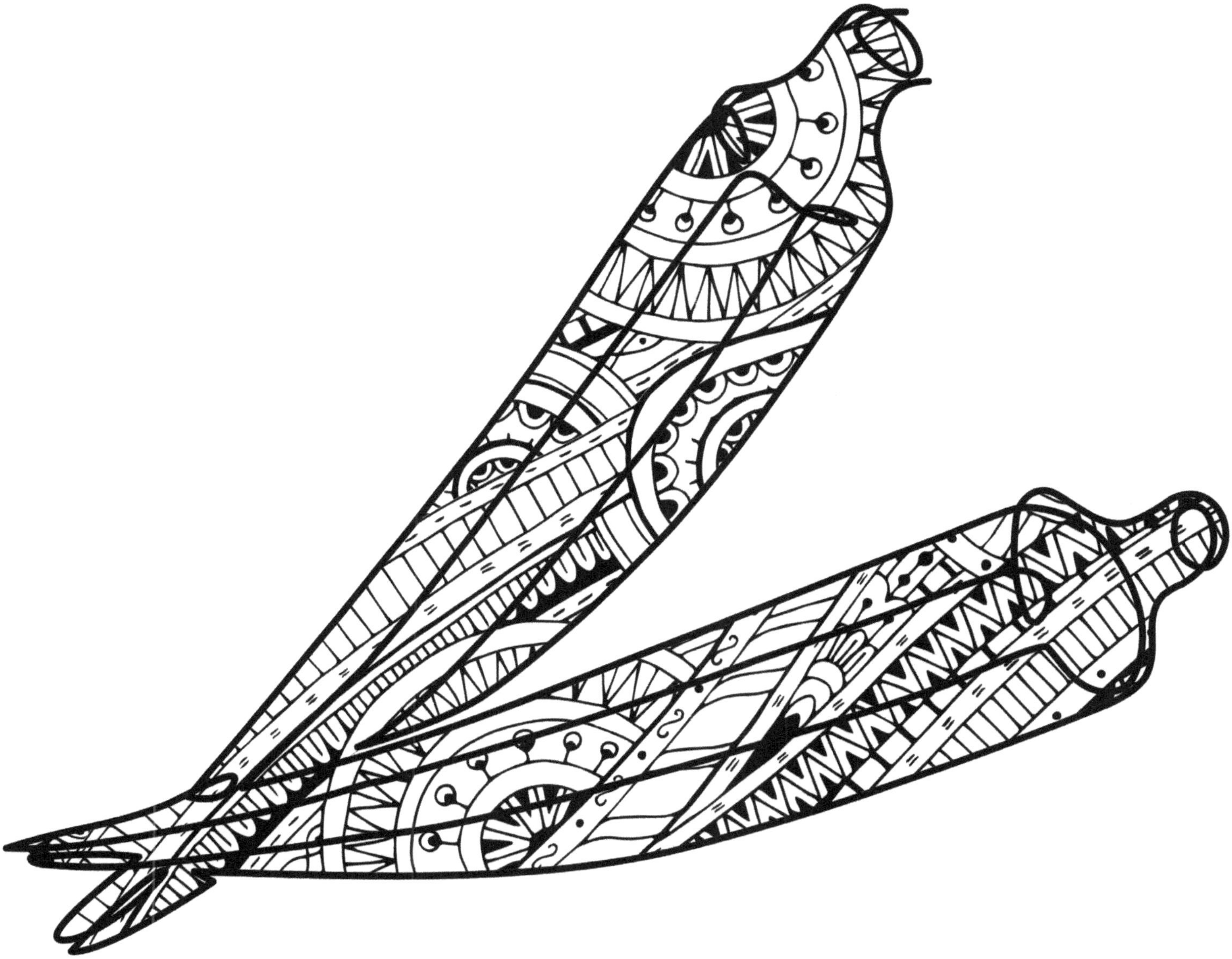

Okra

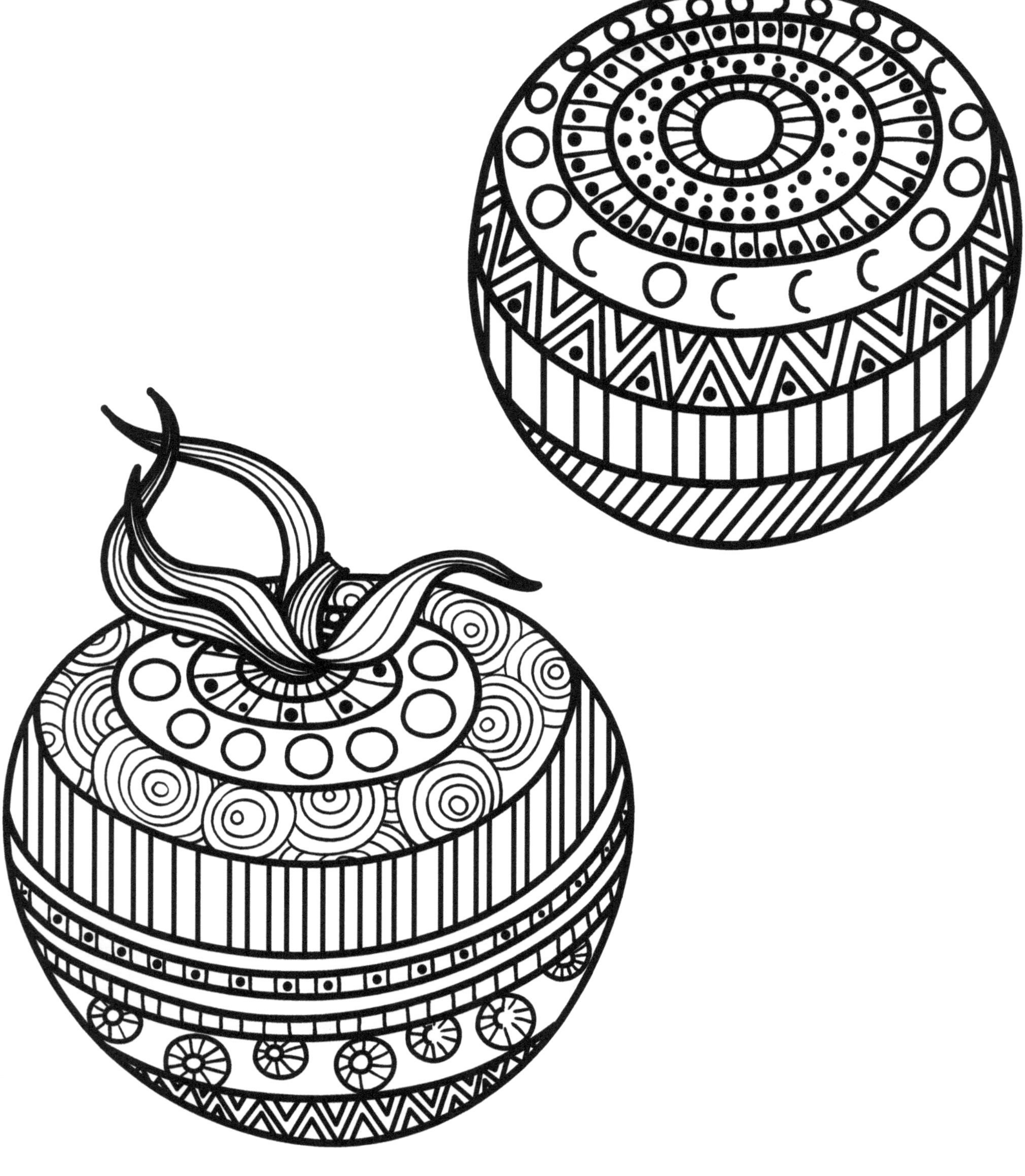

Onions

Onions

Allium cepa

Description

Onions are bulb vegetables which are closely related to garlic, leeks, and chives. There are three different types of onions: red, yellow, and white. All three types of onions have a different taste, and are all used in different culinary dishes. Onions are day-length sensitive. There are "long-day", "intermediate-day", and "short-day" onion types, all of which grow at different times, depending on the length of daylight they are receiving. They are a cold-season crop, and hot temperatures or other stressful conditions cause them to "bolt".

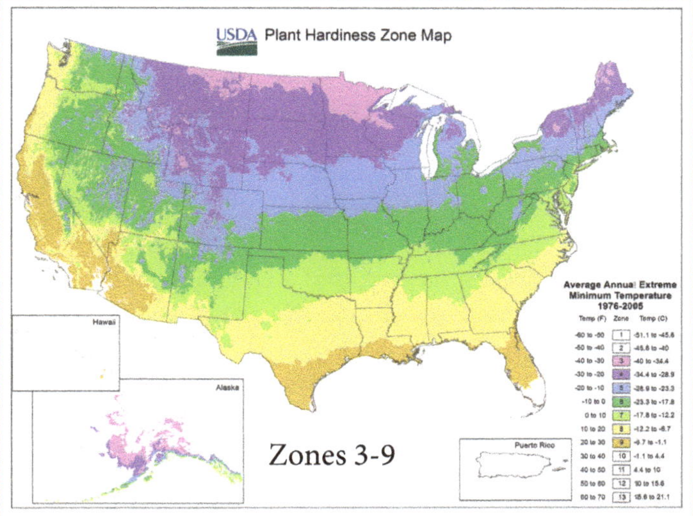

Seeding Square Template

Onion seedlings require 4" spacing, therefore, they are 9 plants per square

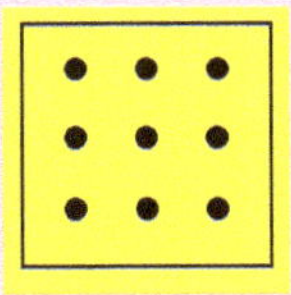

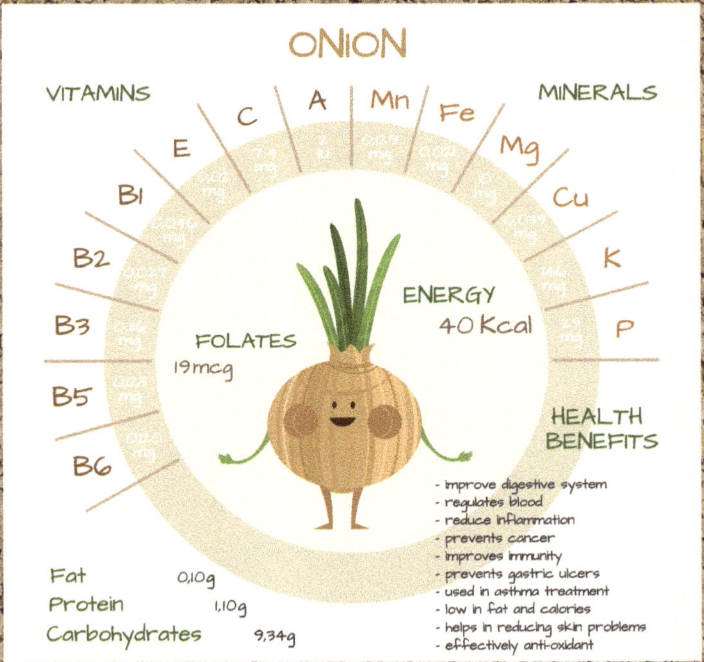

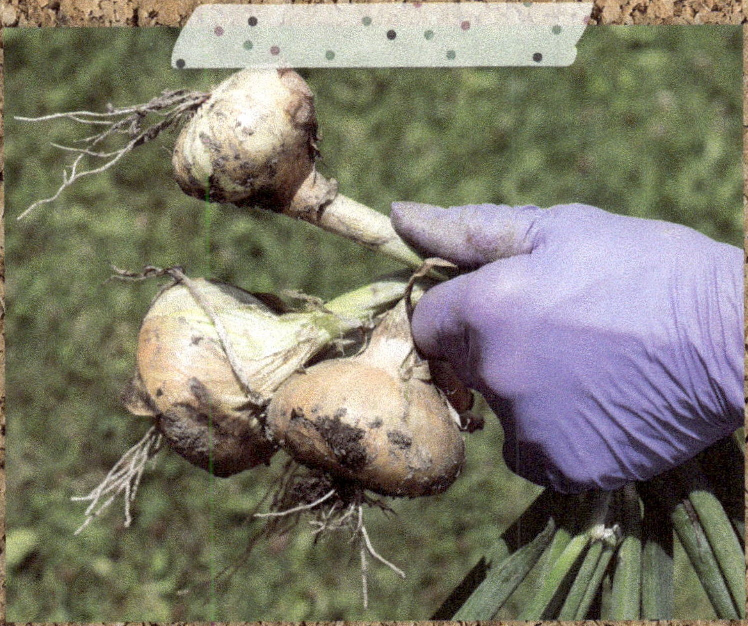

Times to Plant

Onions are a hardy plant, and can be planted while it is still cold outside in the early spring. They can be planted once you are sure your outdoor temperatures do not dip below 20°.

Companion Plants

Beets, strawberry, tomato, lettuce, summer savory, chamomile, roses

Peas

Pisum sativum

Description

The pea is a pod-shaped vegetable, widely grown as a cool season vegetable crop. They do not thrive in the summer heat of warmer temperate and lowland tropical climates, but do grow well in cooler, high altitude, tropical areas. Many cultivars reach maturity about 60 days after planting.

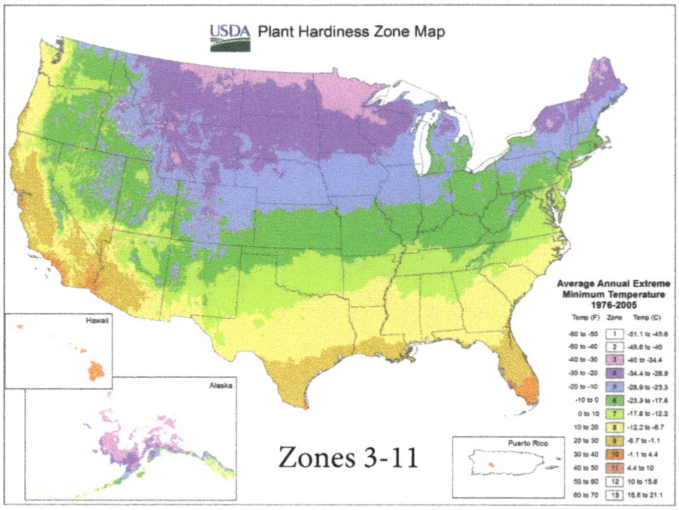

Seeding Square Template

9 plants per square foot

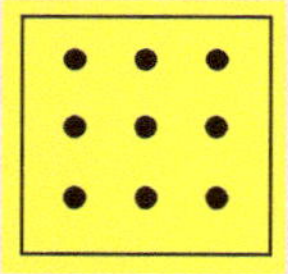

PEAS

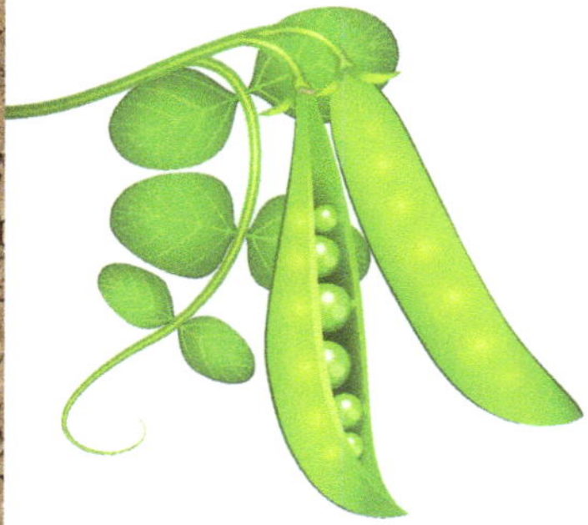

Nutrition Facts

Serving Size 100 grams (100 grams)

Amount Per Serving	
Calories 77	Calories from Fat 3
	% Daily Value*
Total Fat 0g	1%
Saturated Fat 0g	0%
Trans Fat 0g	
Cholesterol 0mg	0%
Sodium 108mg	4%
Total Carbohydrates 14g	5%
Dietary Fiber 5g	18%
Sugars 5g	
Protein 5g	

Vitamin A	41%	Vitamin C	30%
Calcium	2%	Iron	9%

*Percent Daily Values are based on a 2,000 calorie diet. Your daily values may be higher or lower depending on your calorie needs:

	Calories	2,000	2,500
Total Fat	Less than	65g	80g
Sat Fat	Less than	20g	25g
Cholesterol	Less than	300mg	300mg
Sodium	Less than	2,400mg	2,400mg
Total Carbohydrate		300g	375g
Fiber		25g	30g

Calories per gram:
Fat 9 • Carbohydrate 4 • Protein 4

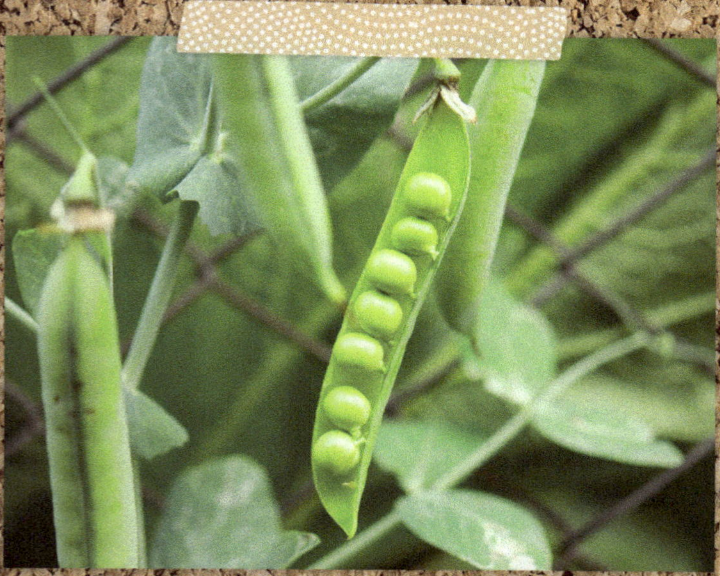

Times to Plant

Till your garden for peas in the early fall, but sow your seeds directly in your garden 4-6 weeks before your final frost, when soil temperatures reach 45°.

Companion Plants

Beans, Carrots, Corn, Cucumbers, Eggplant, Lettuce, Melons, Parsnips, Potatoes, Radishes, Spinach, Turnips

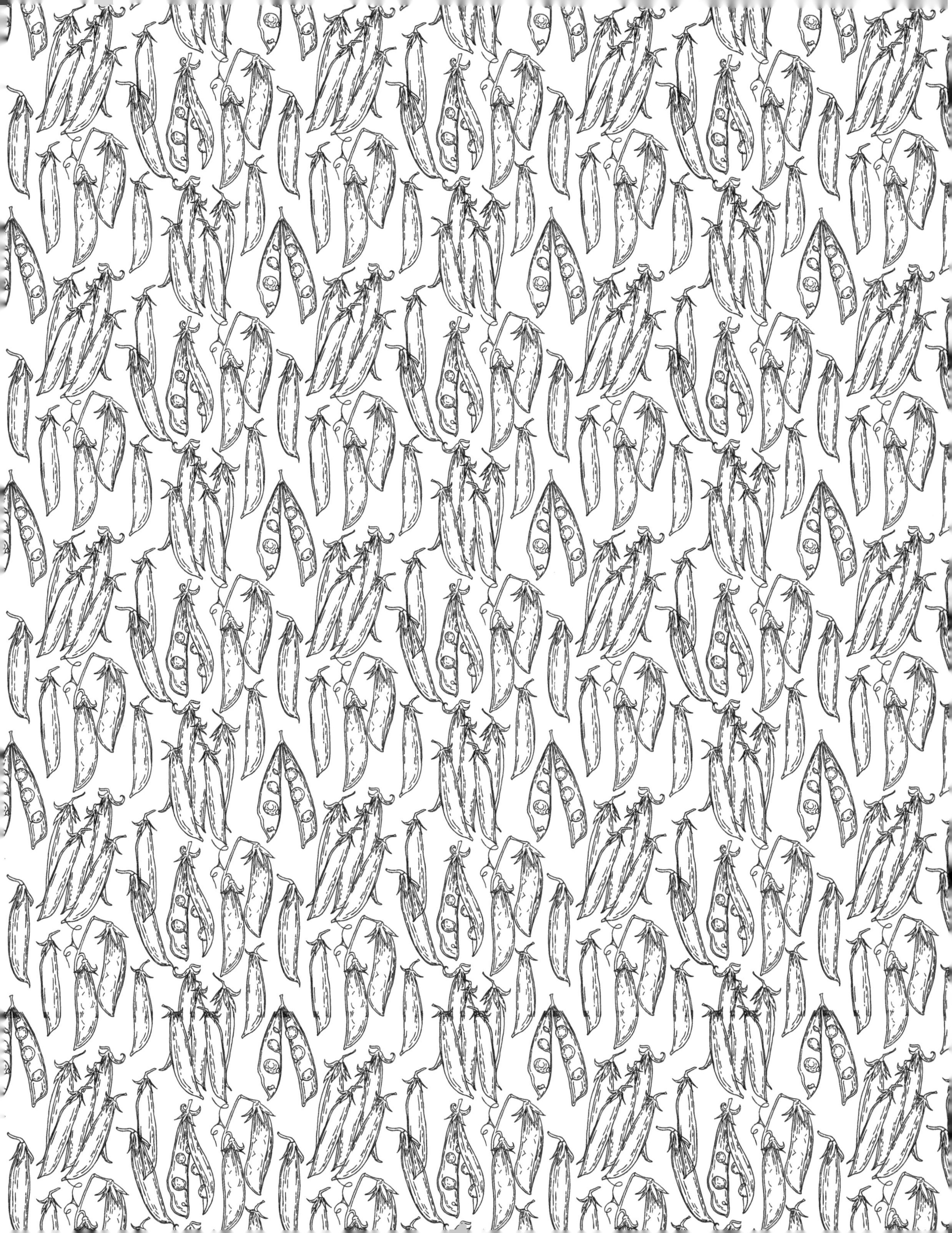

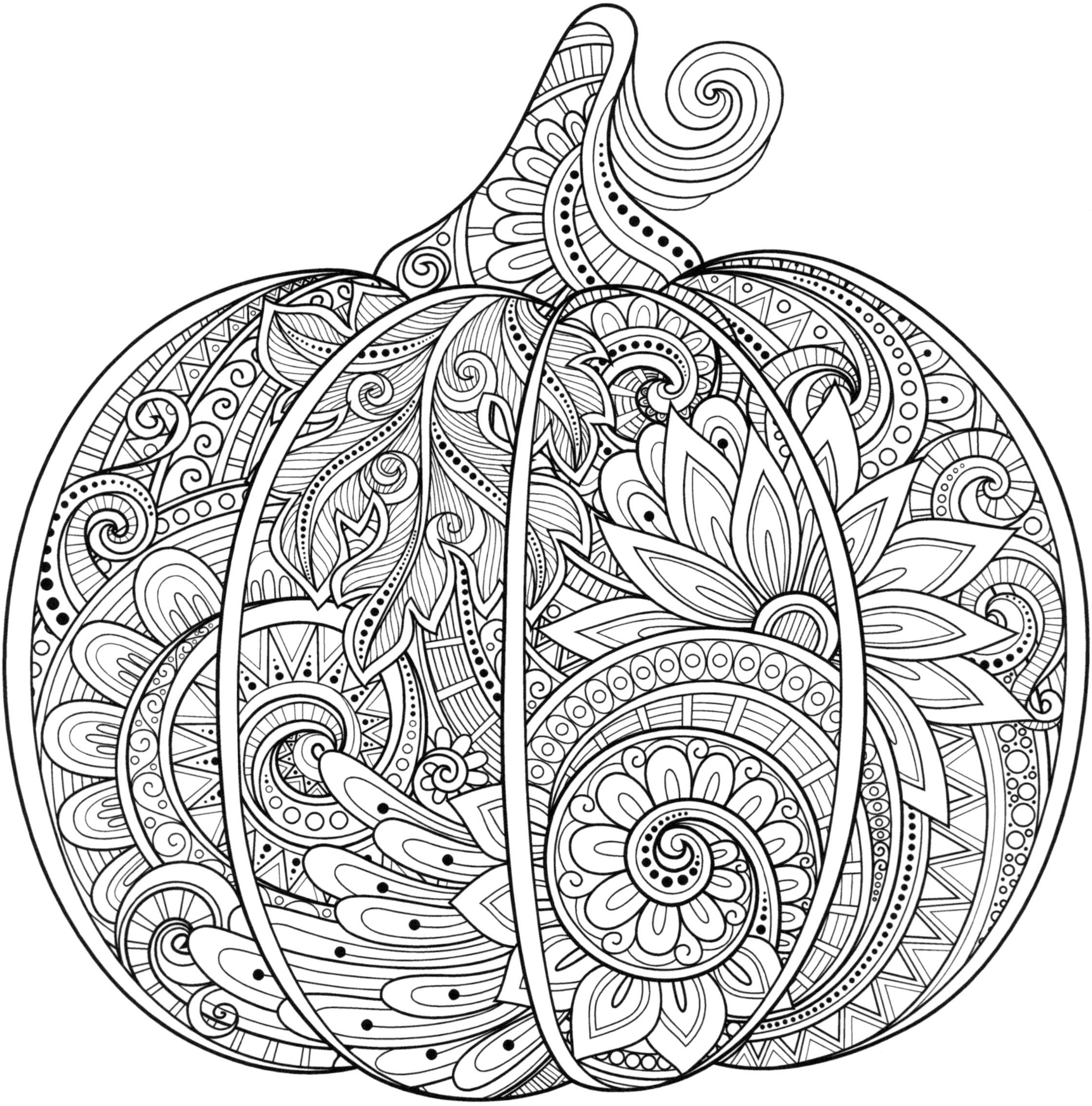

Pumpkin

Pumpkin

Cucurbita pepo

Description

Part of the winter squash family, pumpkins are a vine vegetable that crave lots of warmth and sunlight. They produce both a male and female flower, and honeybees play a significant role in fertilization. Pumpkins have many health benefits, and are perfect for the cooler months when we tend to get sick.

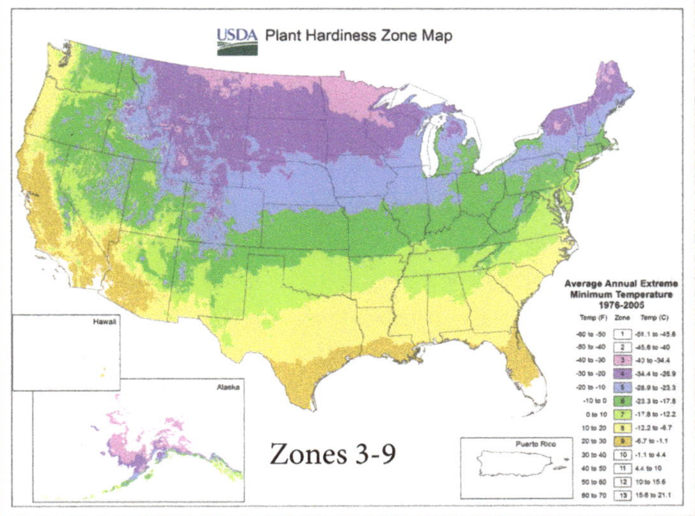

Seeding Square Template

4+ square feet per plant

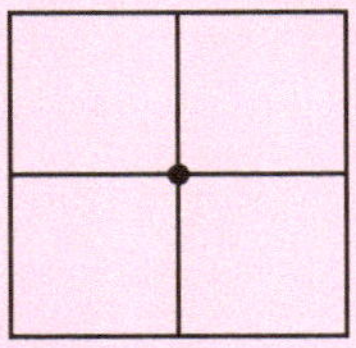

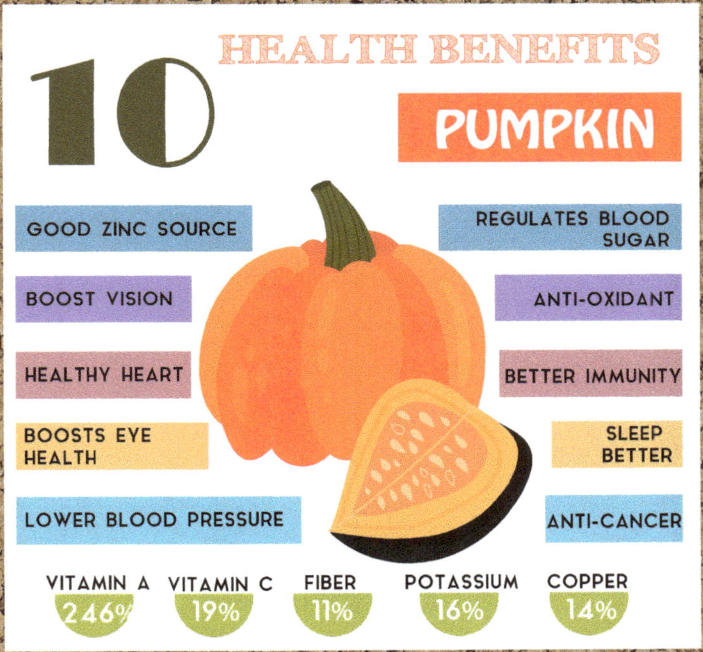

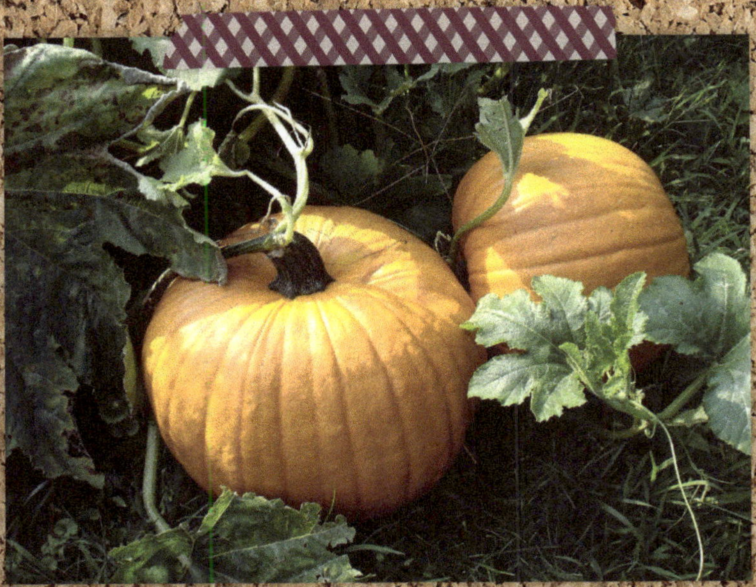

Times to Plant

Ideal soil temperature for pumpkin seeds is 95°, so wait until your soil temperatures are a minimum of 70° before directly planting seeds in your garden. They require at least 100 days of frost-free growing.

Companion Plants

Datura, corn, pole beans

Radish

Raphanus raphanistrum

Description

The radish is an edible root vegetable, grown and consumed throughout the world, being mostly eaten raw as a crunchy salad vegetable. Radishes owe their sharp flavor to the various chemical compounds produced by the plants, including glucosinolate, myrosinase, and isothiocyanate. They are sometimes grown as companion plants and suffer from few pests and diseases.

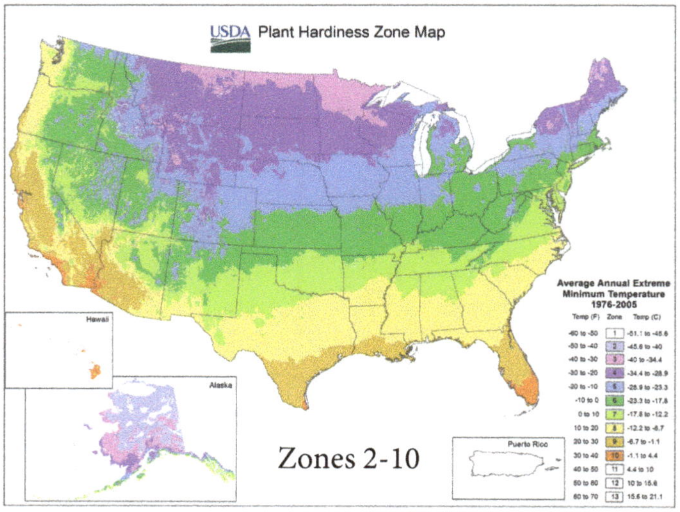

Seeding Square Template

3" per seed, or 16 seeds per square

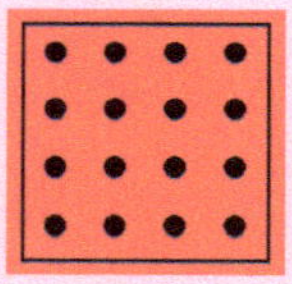

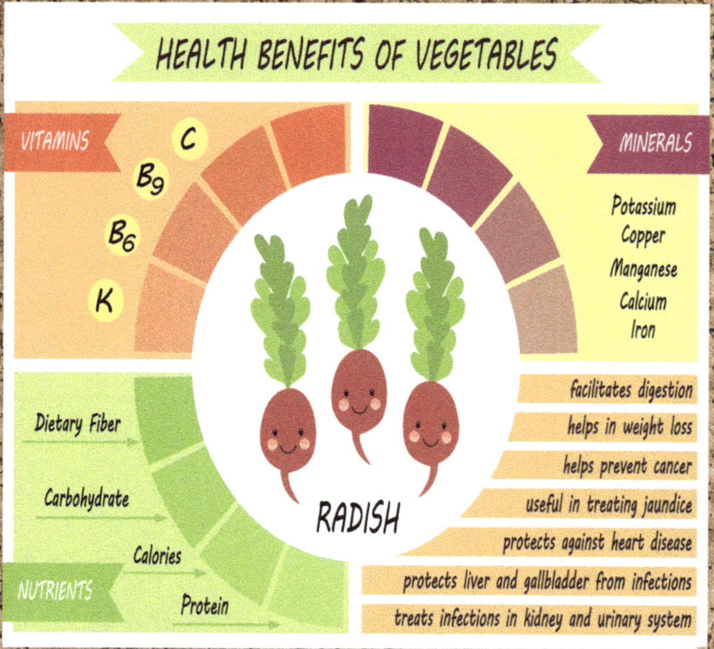

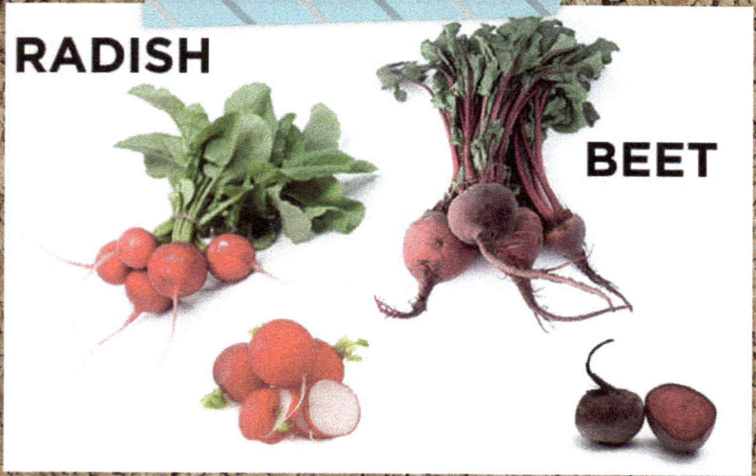

Times to Plant

Sow seeds in a sunny spot approximately 4-6 weeks before your final frost date.

Companion Plants

Peas, nasturtium, lettuce, cucumber, beets, spinach, carrots, squash, melons, tomatoes, beans

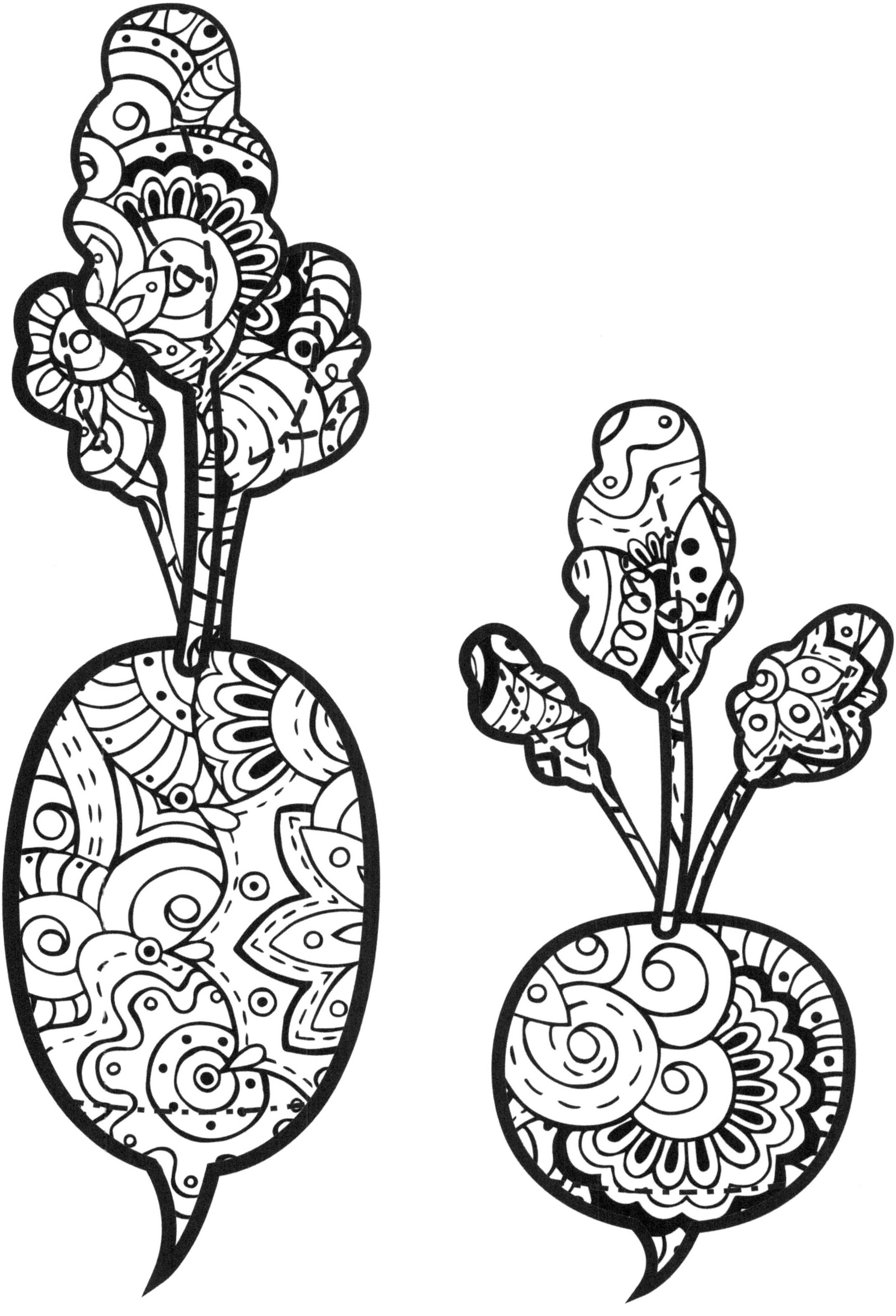

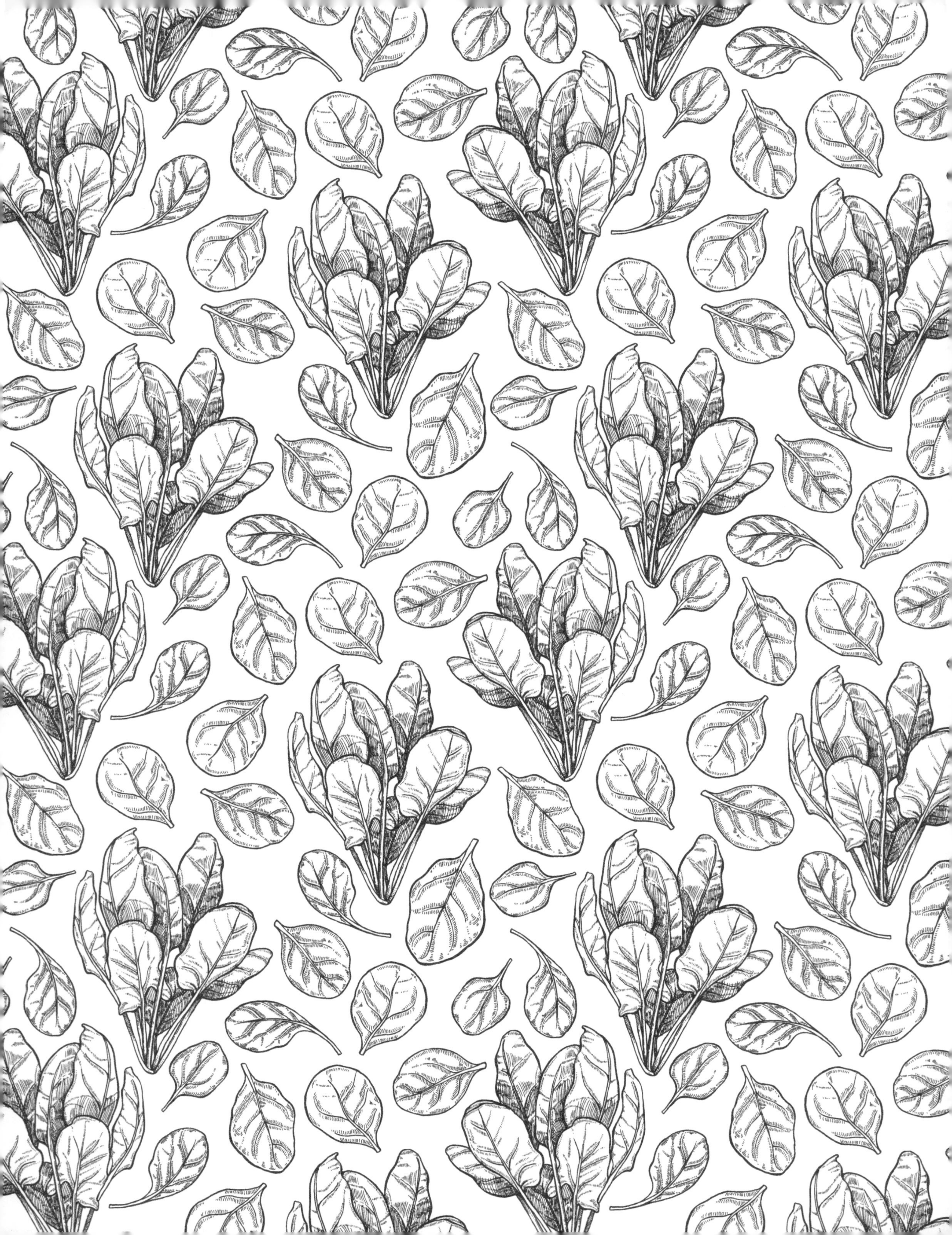

Spinach

Spinacia oleracea

Description

Unlike lettuce, Spinach is a part of the Amaranthaceae family, and is an edible flowering plant. Its leaves are commonly eaten as a vegetable, either fresh, frozen, canned, chipped, or dehydrated. It is a very cold-hardy vegetable, so it can be planted in very early spring, as well as fall and even winter. Spinach is one of the highest scoring vegetables when it comes to iron, calcium, and other vitamins, and is one of the best sources of vitamins A, B, and C.

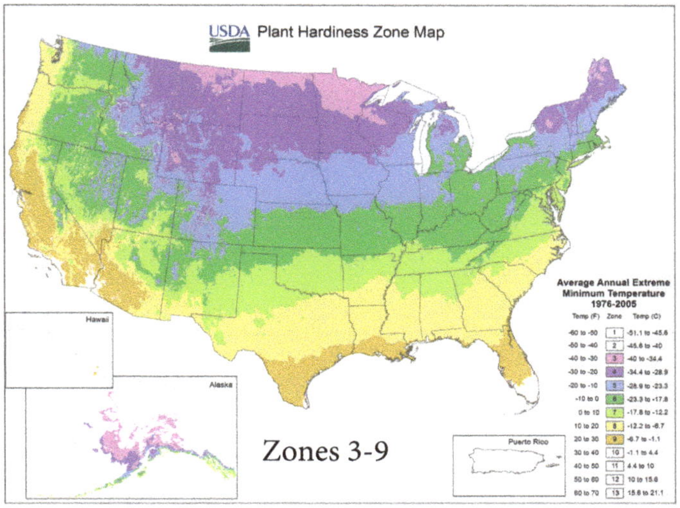

Seeding Square Template

4" per plant, so 9 plants per square foot

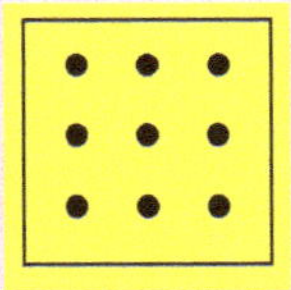

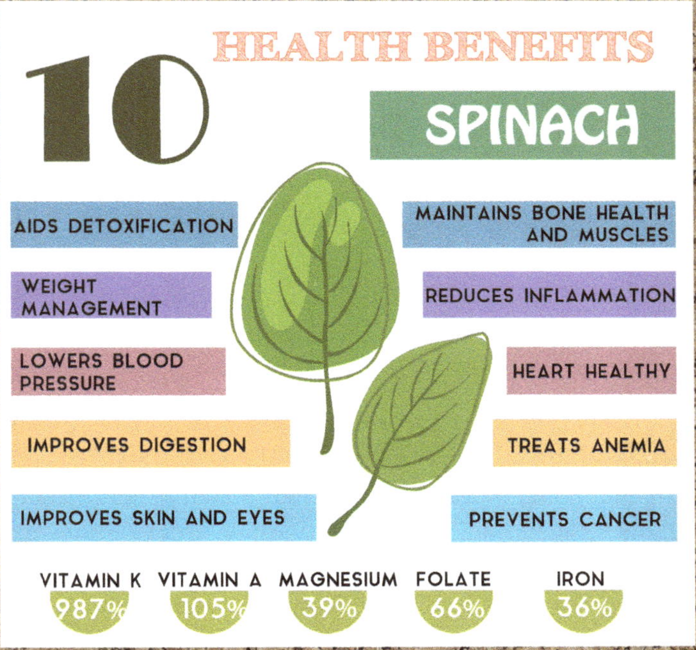

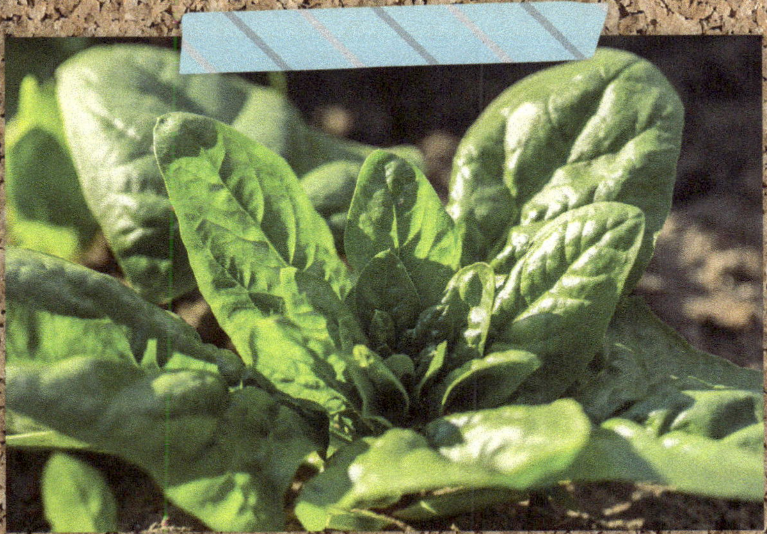

Times to Plant

Spinach can be planted directly in your garden approximately 5 weeks before your final frost date.

Companion Plants

Asparagus, Brussels, Cabbage, Celery, Dill, Eggplant, Lettuce, Onions, Peas, Peppers, Radishes, Strawberries, Tomatoes

Sweet Potato

Ipomoea batatas

Description

Sweet potatoes are root vegetables that have vines that run above ground. They originated in Central and South America, so they prefer warm, humid, sunny climates. They need plenty of space, sunlight, and water, but one must be careful not to over-water. They thrive the most in sandy soils. Sweet potatoes are on the list of the top ten healthiest foods, ranking number one when vitamins A and C, iron, calcium, protein, and complex carbohydrates were considered.

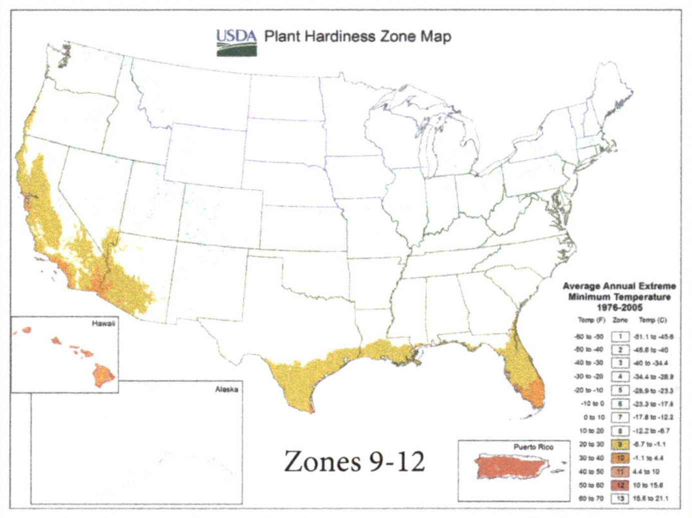

Seeding Square Template

8" deep x 1 sq ft per plant

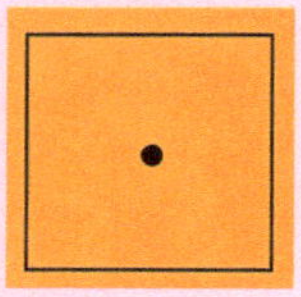

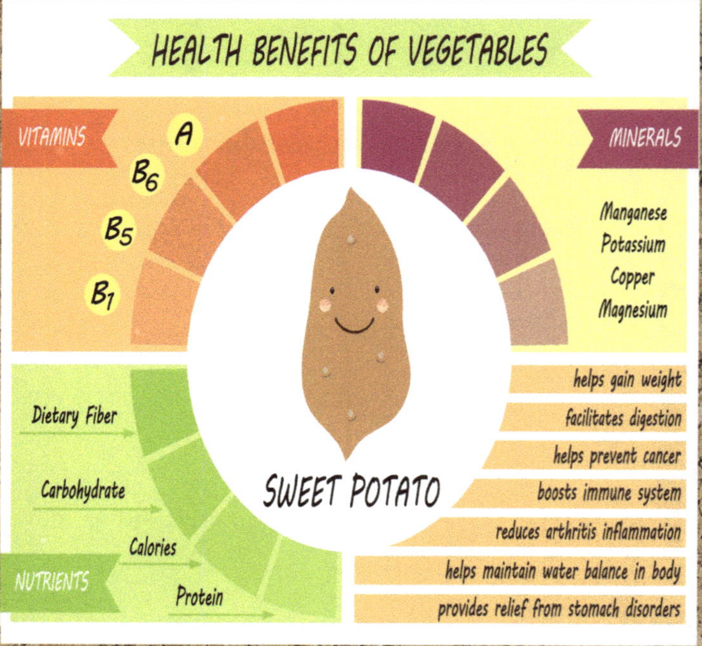

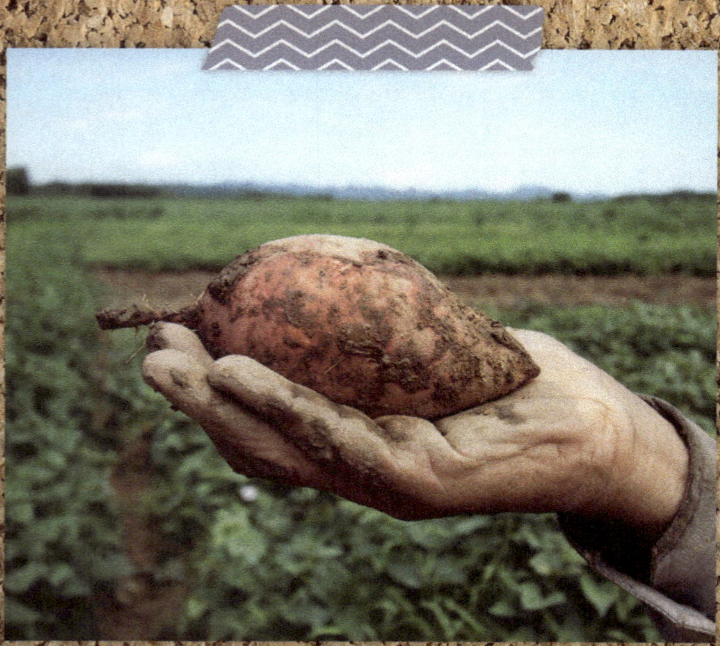

Times to Plant

Sweet potatoes can be grown year-round, but need a minimum of 4 months of warmth and sunshine in order to properly grow.

Companion Plants

Root vegetables such as beets and parsnips. Bush beans, potatoes, thyme, oregano, and dill.

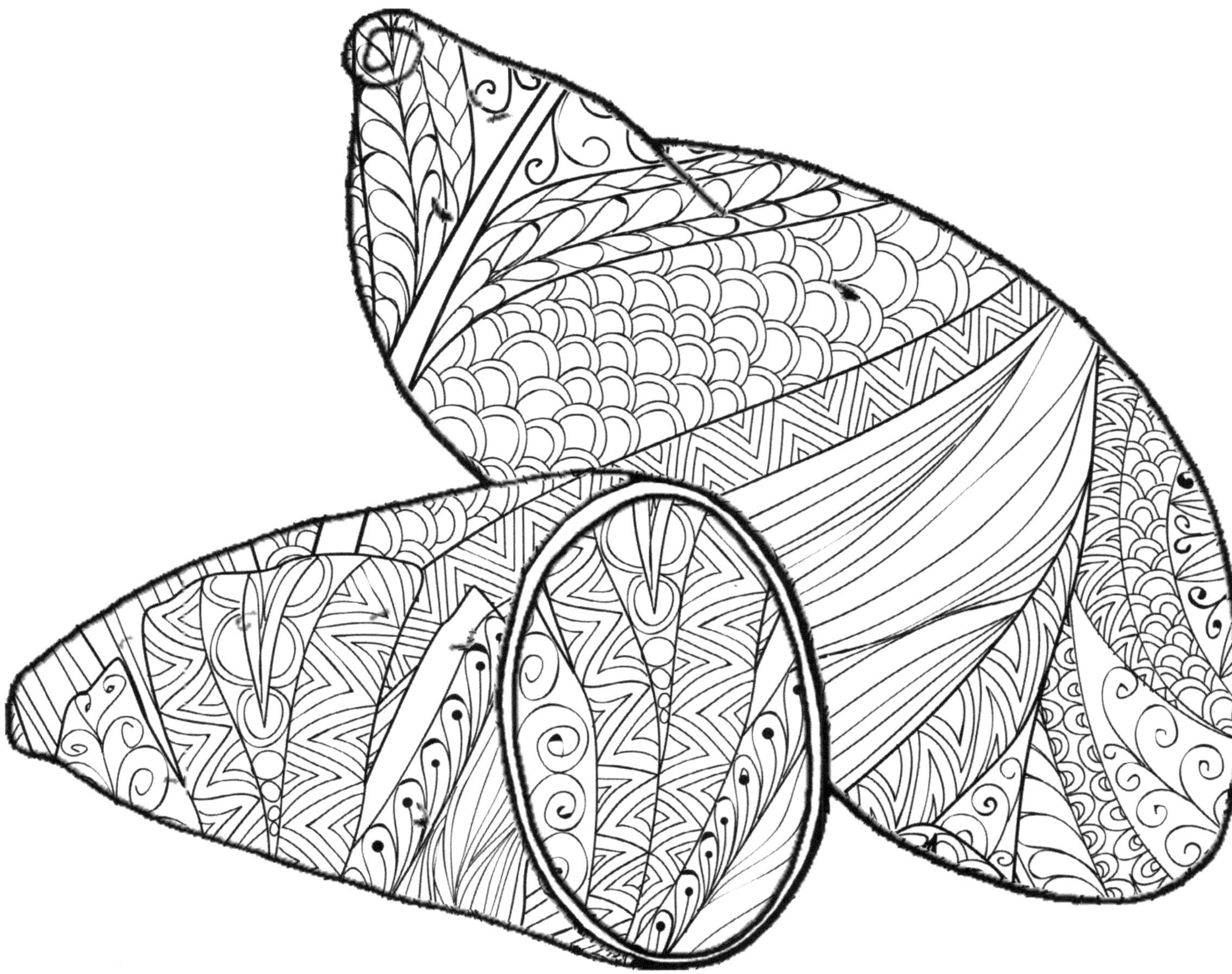

Sweet Potato

Tomato

Solanum lycopersicum

Description

Originating in western South America, the tomato plant belongs to the nightshade family. There are around 7,500 tomato varieties known. There is quite the debate as to whether or not tomatoes are fruits or vegetables, because a tomato is a fruit, a berry, consisting of the ovary, together with its seeds, all of a flowering plant. However, it could also be considered a vegetable, because it has a much lower sugar content than culinary fruits. The tomato plant is often considered to be a good plant for beginner gardeners.

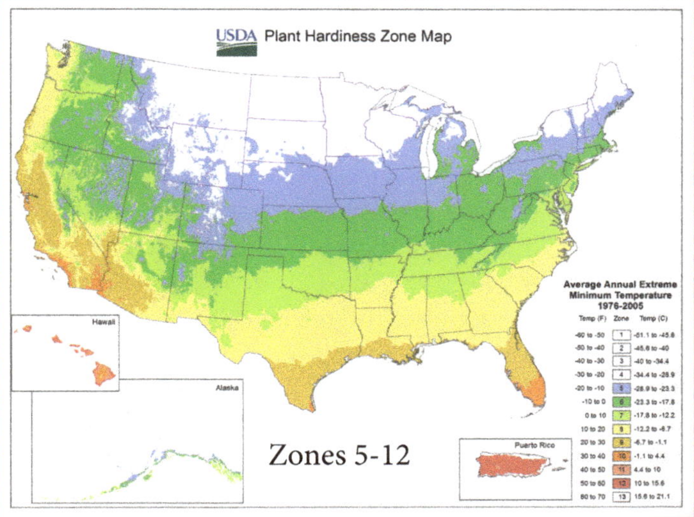

Seeding Square Template

Vining tomatoes require 1 sq ft per plant, however, bush tomatoes require 4 sq ft per plant.

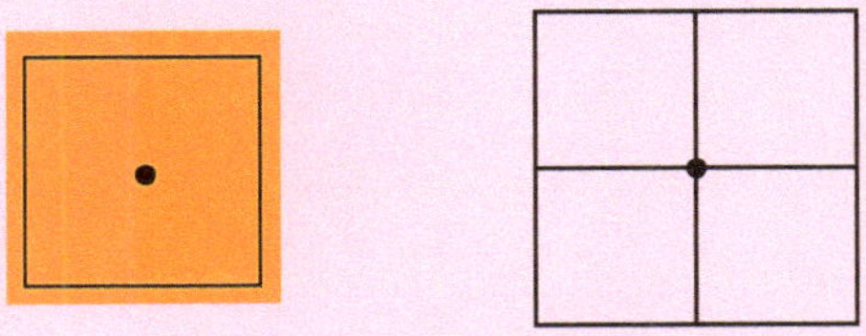

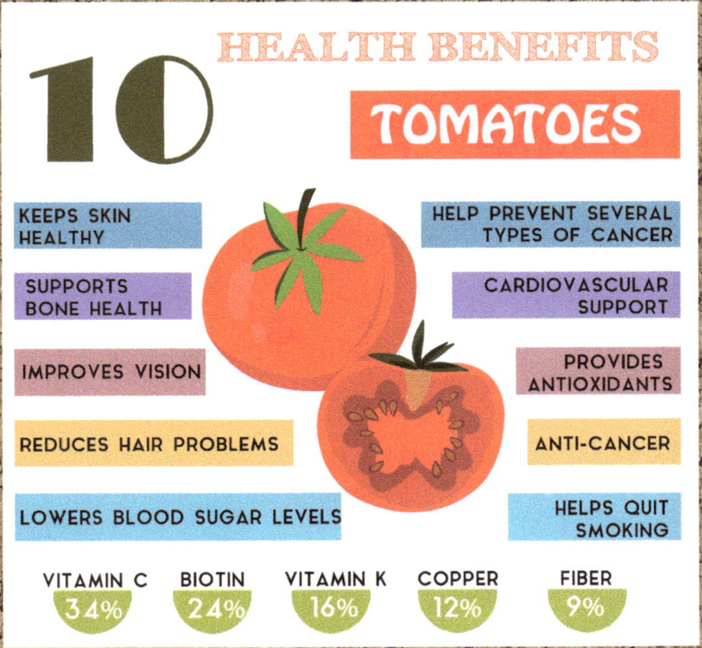

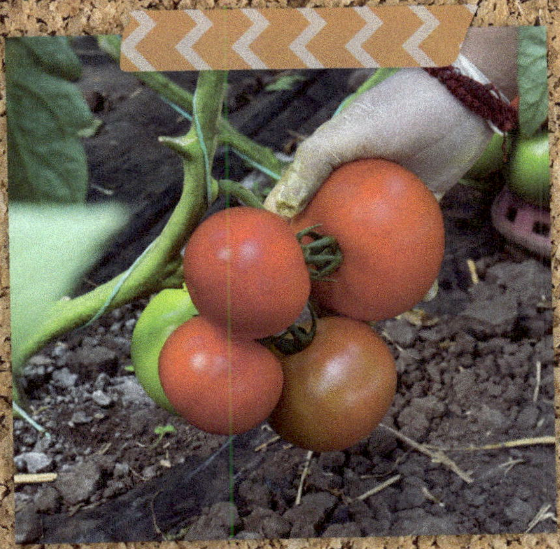

Times to Plant

In zones 8 and 9, start seeds as early as mid-January; in zones 3 or 4, wait to start seeds until mid- to late March and early April. Then wait two weeks after the last frost date to plant seeds or set transplants outdoors.

Companion Plants

Onions, chives, garlic, mints, French marigold, asparagus, carrots

Turnip

Brassica rapa

Description

Like lettuce, turnips are a vegetable that grows rapidly in your garden. They are a root vegetable, which produces a bulbous root, as well as greenery. The greens are cooked and are known as turnip greens. While they are consumed by humans in the United States, they are commonly used as feed for livestock in England, Scotland, Ireland, and other European countries. There have also been hybrid plants that have been made with turnips, such as the Rutabaga.

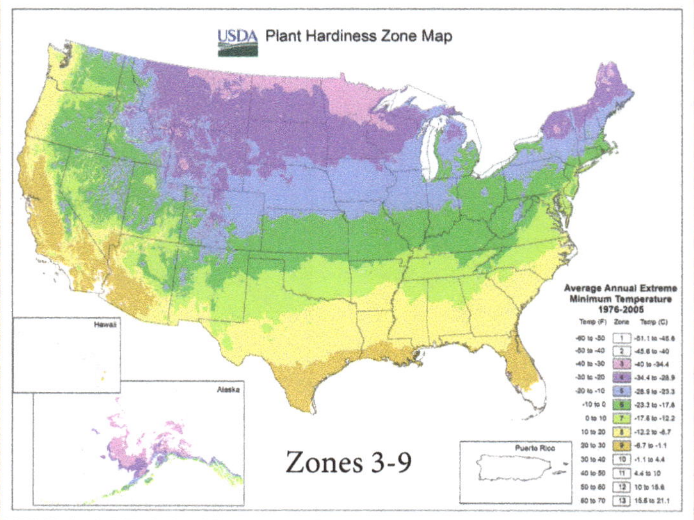

Seeding Square Template

4" per turnip, so 9 plants per square

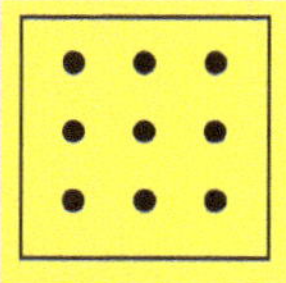

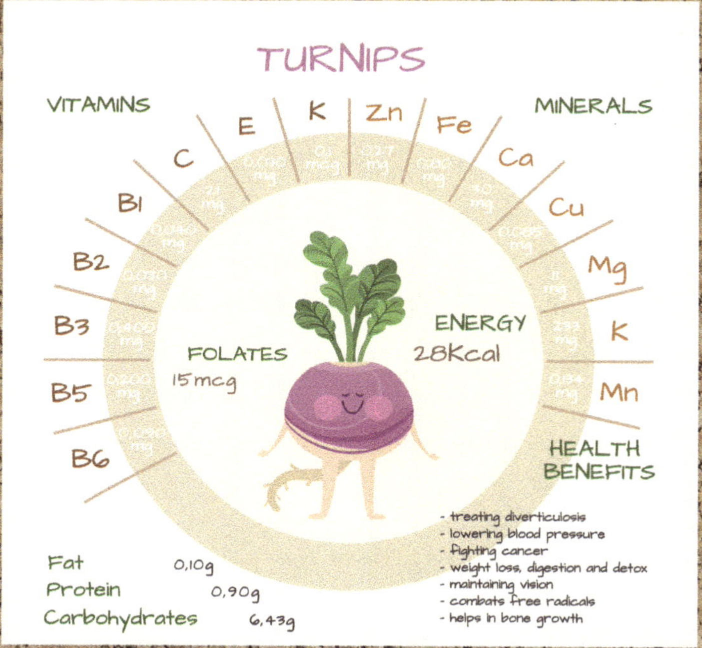

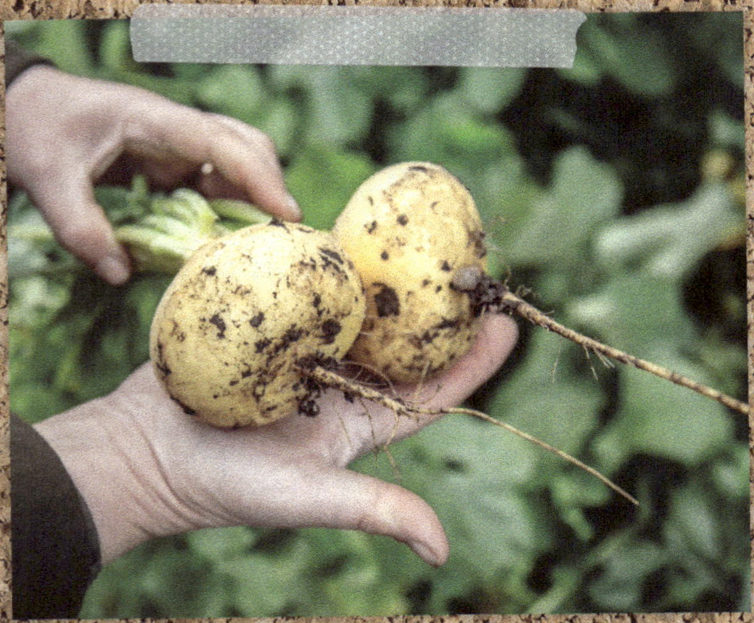

Times to Plant

Plant directly in your garden about 2 to 3 weeks before your final frost date. For an autumn harvest, you can also plant in late summer.

Companion Plants

Peas, vetch, mint, rosemary, sage

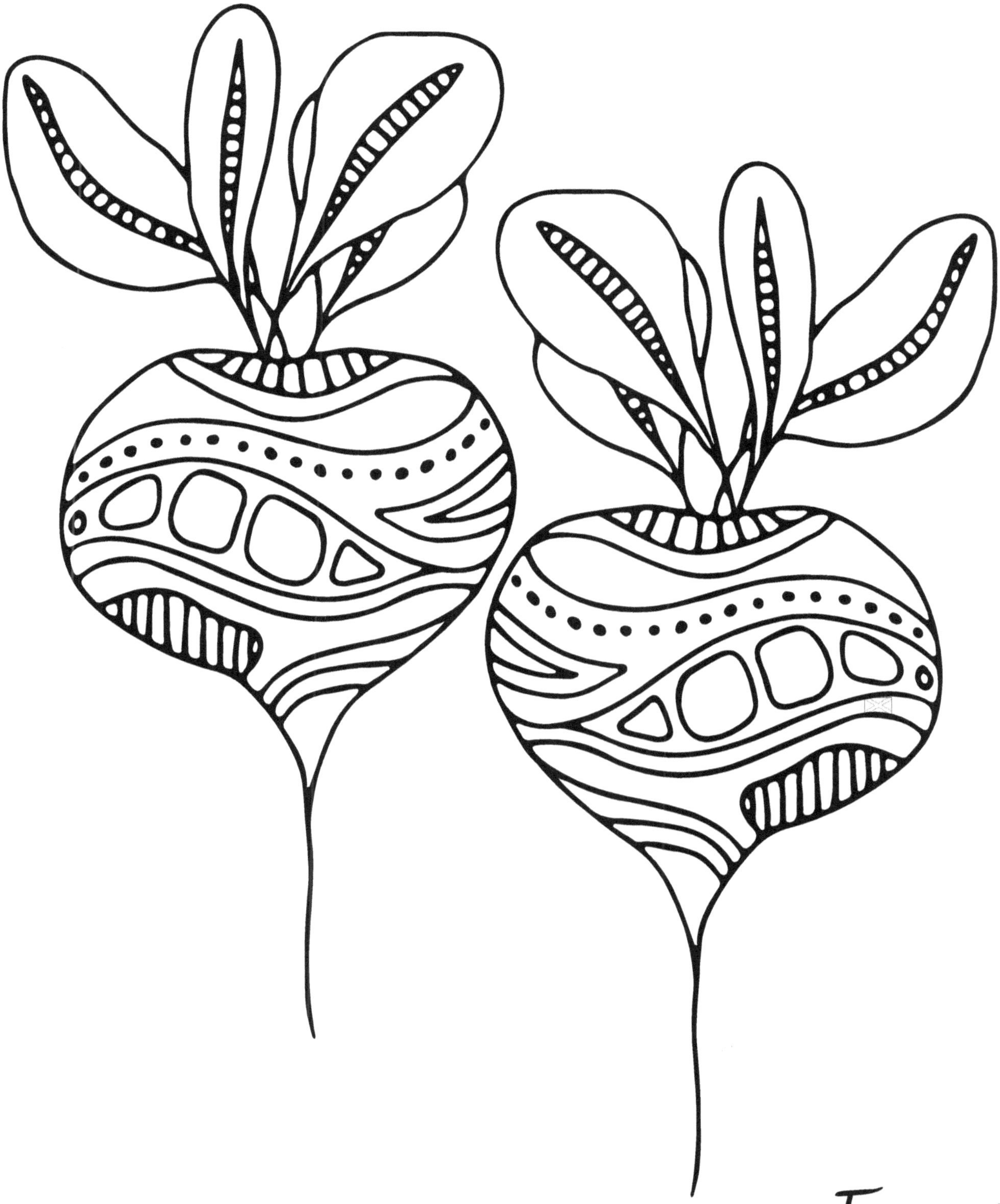

Turnips

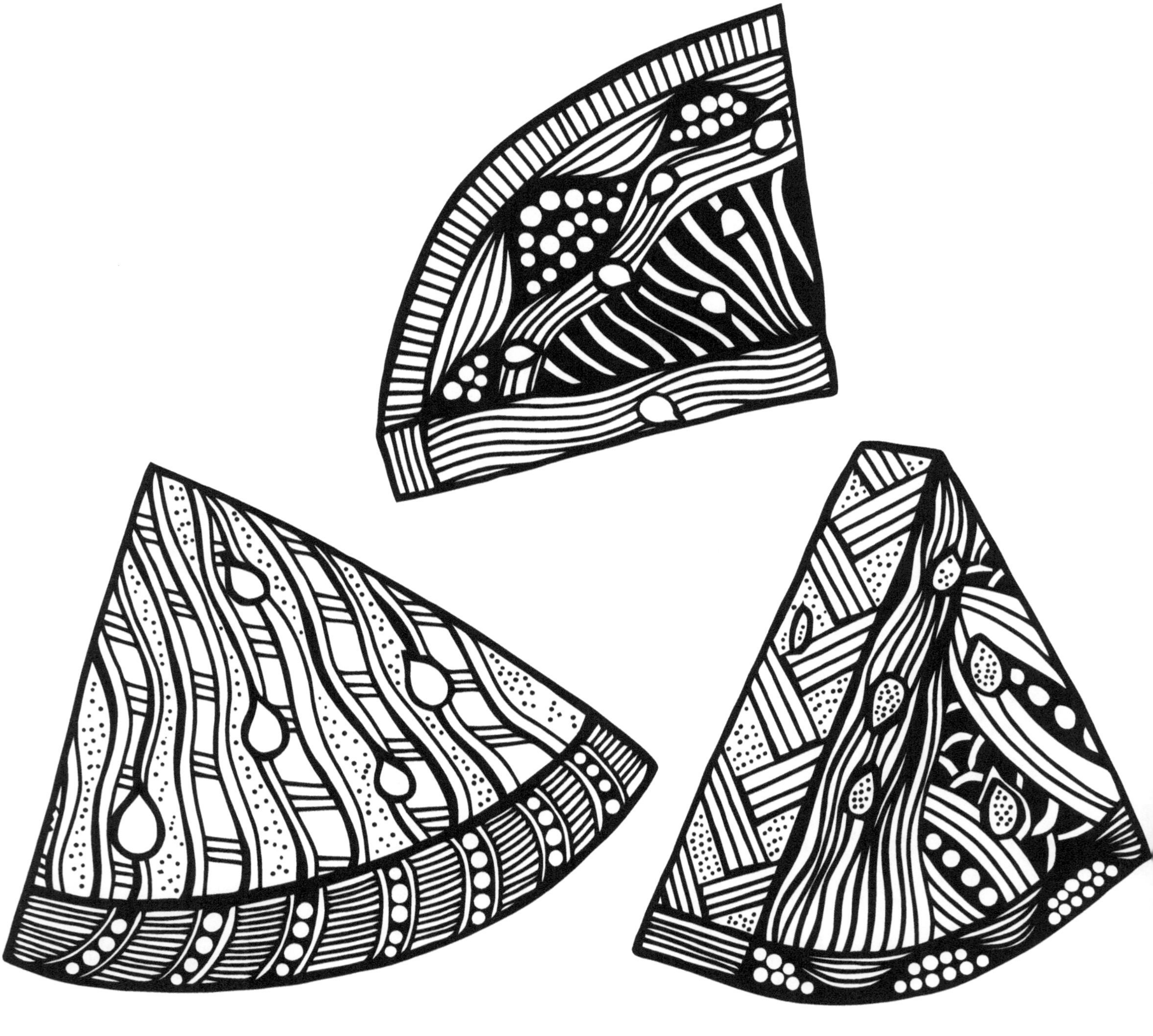

Watermelon

Watermelon

Citrullus lanatus

Description

Watermelon is a scrambling and trailing vine in the flowering plant family, with considerable breeding efforts being put into disease-resistant varieties. Many cultivars are available that produce mature fruit within 100 days of planting the crop.

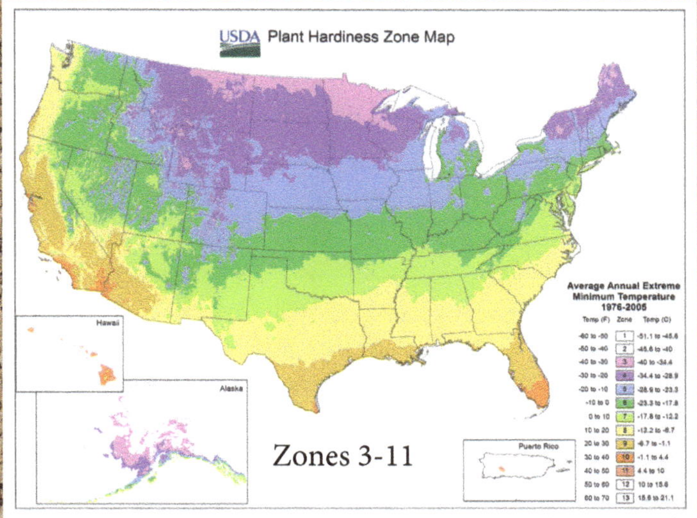

Seeding Square Template

4+ square feet per plant

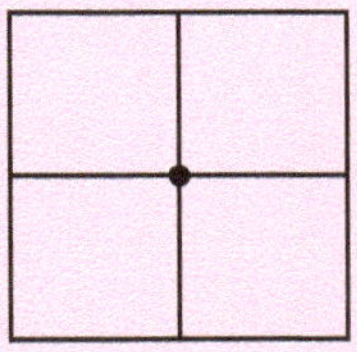

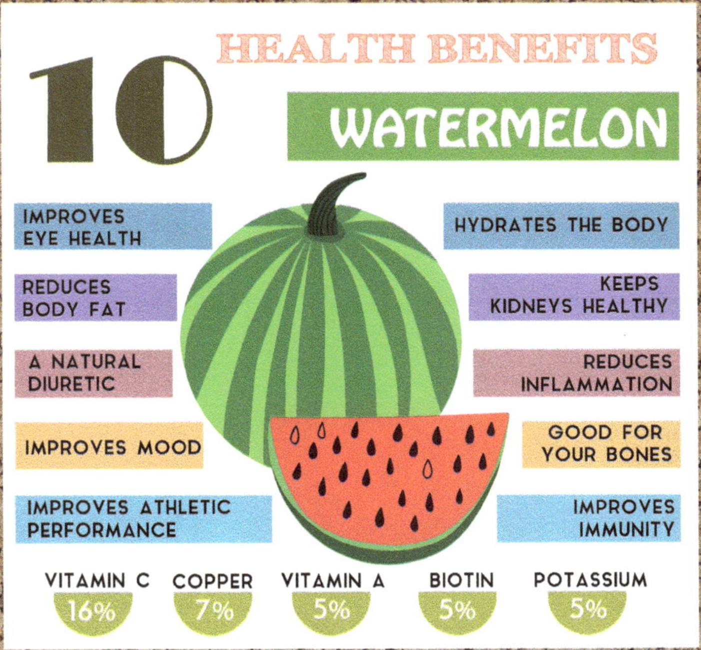

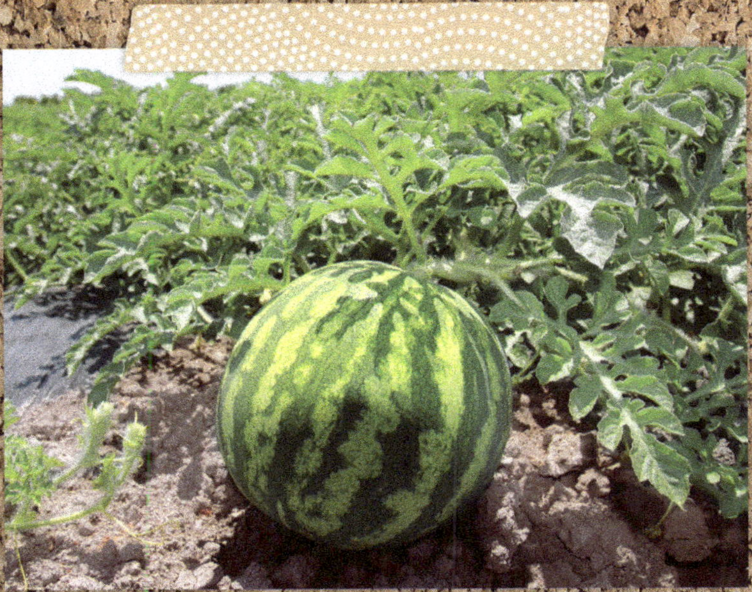

Times to Plant

Sow seeds directly into your garden once your soil temperature reaches a consistent 70°, or start seeds indoors and then transplant the vines in your garden two weeks after your final frost date.

Companion Plants

Marigolds, radishes, spearmint, oregano, basil, corn

Yellow Squash

Cucurbita pepo

Description

Yellow squash are a buttery yellow, elongated squash that are harvested while they are still immature. They are a type of summer squash, due to the fact that are harvested early, as compared to the winter squash. If they are left too long, they become warted. The plants of summer squash are bushes, rather than vines, therefore, the Yellow Squash is found on bushes.

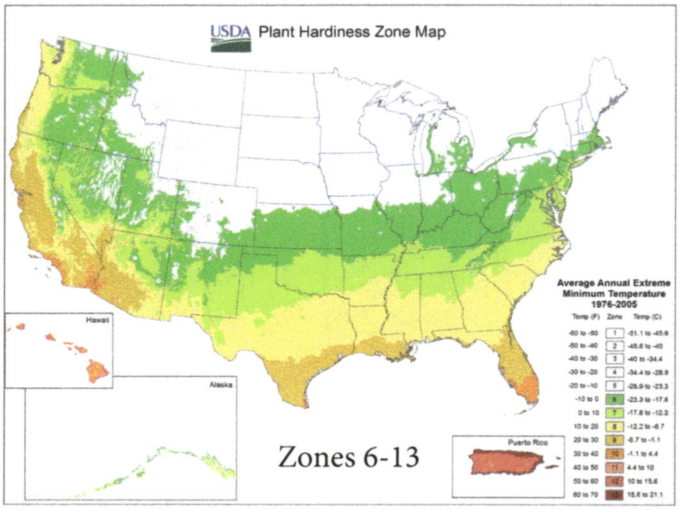

Seeding Square Template

4+ square feet per plant

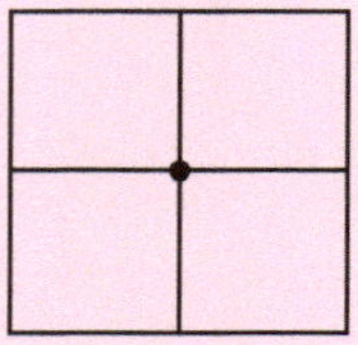

YELLOW SQUASH

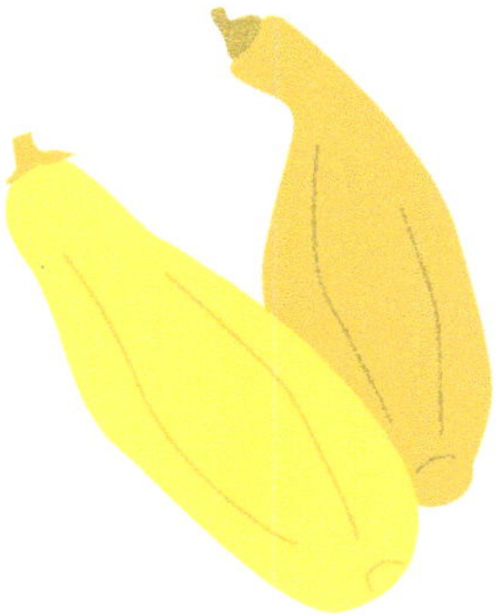

Nutrition Facts

Serving Size ½ medium squash (98g)

Amount Per Serving	
Calories 20	Calories from Fat 0
	% Daily Value*
Total Fat 0g	0%
Saturated Fat 0g	0%
Trans Fat 0g	0%
Cholesterol 0mg	0%
Sodium 0mg	0%
Total Carbohydrate 4g	1%
Dietary Fiber 2g	8%
Sugars 2g	
Protein 1g	

Vitamin A 6%	Vitamin C 30%
Calcium 2%	Iron 2%

* Percent Daily Values are based on a 2,000 calorie diet. Your daily values may be higher or lower depending on your calorie needs:

	Calories	2,000	2,500
Total Fat	Less Than	65g	80g
Saturated Fat	Less Than	20g	25g
Cholesterol	Less Than	300mg	300mg
Sodium	Less Than	2,400mg	2,400mg
Total Carbohydrate		300g	375g
Dietary Fiber		25g	30g

Calories per gram:
Fat 9 Carbohydrate 4 Protein 4

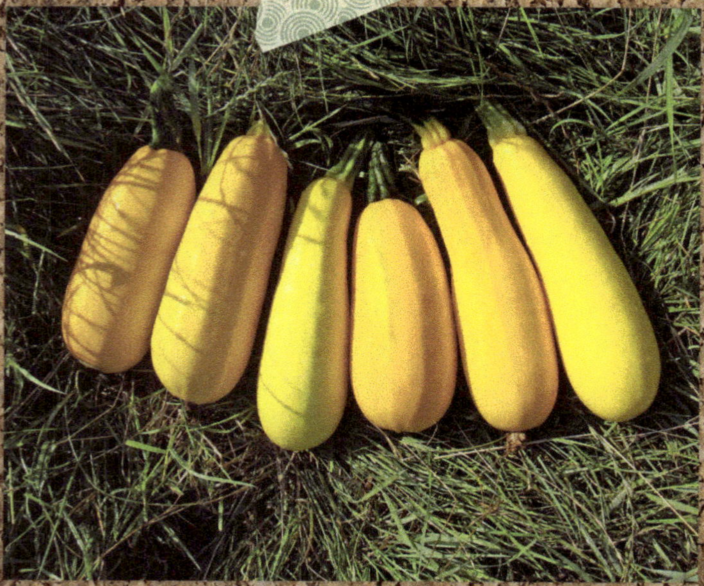

Times to Plant

After the final frost date in your zone, through the twelfth week before your average first fall frost date.

Companion Plants

Legumes, radishes, dill, corn, and oregano

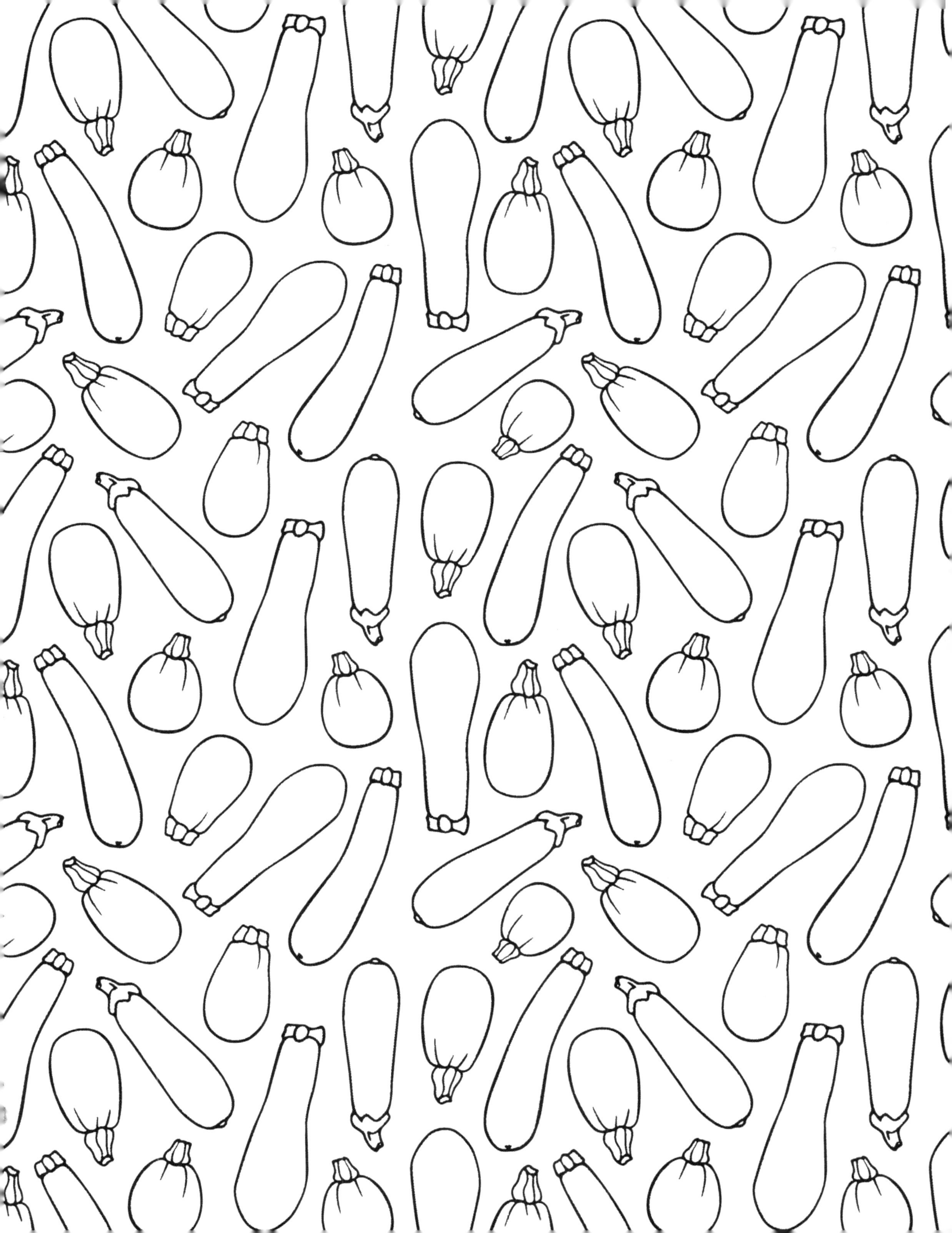

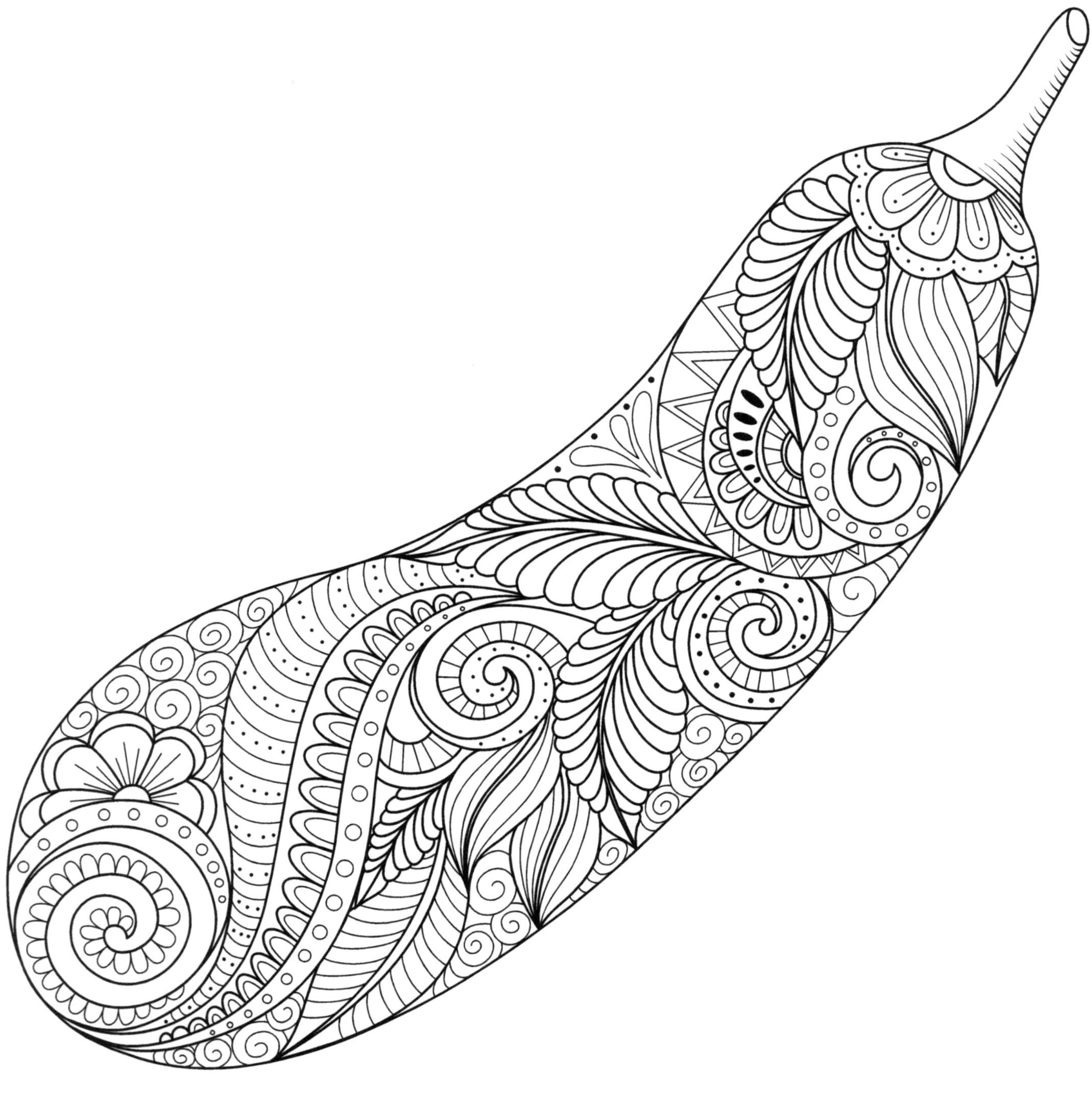

Zucchini

Zucchini

Cucurbita pepo

Description

Another type of summer squash, Zucchini grow on bushes. They are an elongated, green gourd, similar in appearance to cucumbers. They are usually harvested while still immature, just like other summer squashes. The gourd is very easy to cultivate in temperate climates. As such, it has a reputation among home gardeners for overwhelming production.

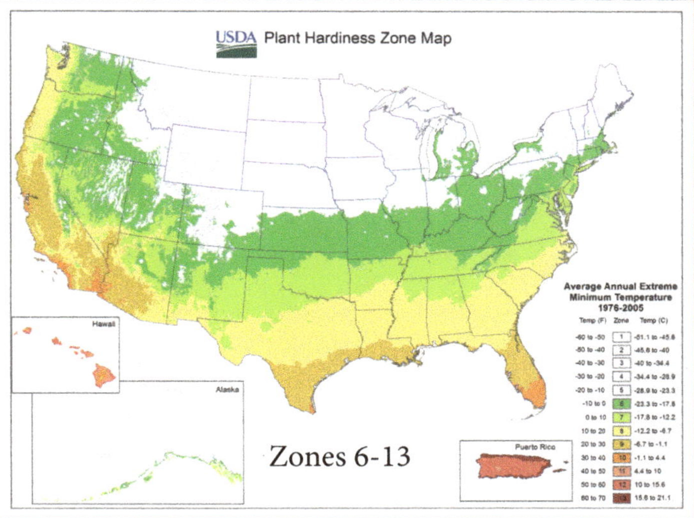

Seeding Square Template

4+ square feet per plant

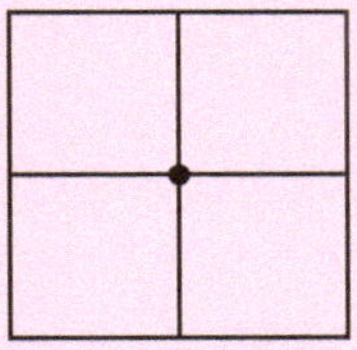

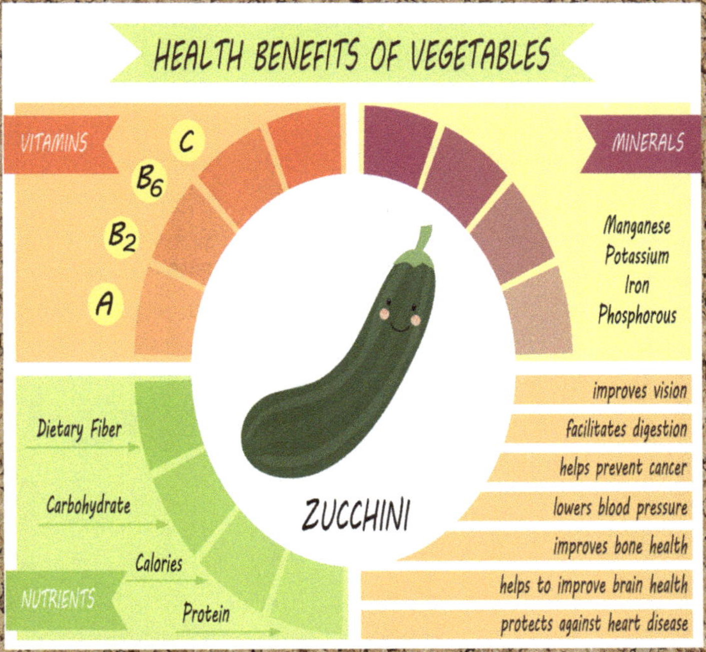

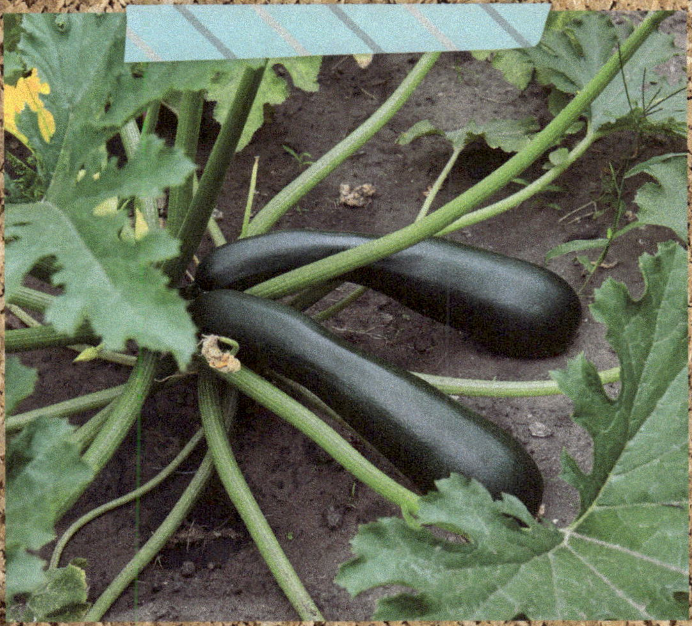

Times to Plant

Plant in about the month of May, however, wait until outdoor temperatures stay reliably above 70°, and your soil temp is in the 60s.

Companion Plants

Parsley, Spinach, Garlic, Corn, Beans, Peas, and Nasturtiums.

Enjoy other books by

LEARN AND COLOR

Current and upcoming titles:

Learn and Color Nature Series

- Medicinal Herbs
- Freshwater Fish
- Garden Edibles
- Trees
- Bugs and Insects
- Fossils

Learn and Color Stained Glass Series

- Landscapes & Seascapes
- Fish & Fowl
- Flowers
- Birds, Bees, & Butterflies
- Animals
- Designs
- Sea Creatures
- Christian Images

Learn and Color the Bible Series

- Promises
- Prayers
- Proverbs
- Places
- Psalms
- People

Learn and Color Historical Figures Series

- Early Civilization
- The Ancient World
- The Middle Ages
- The Renaissance and Reformation
- The Industrial Revolution
- The Modern Age
- The Present Day

Learn and Color Historical Events Series

- Early Civilization
- The Ancient World
- The Middle Ages
- The Renaissance and Reformation
- The Industrial Revolution
- The Modern Age
- The Present Day

Learn and Color Historical Places Series

- Early Civilization, The Ancient World
- The Middle Ages, Renaissance & Reformation
- The Modern Age

www.ingramcontent.com/pod-product-compliance
Lightning Source LLC
Chambersburg PA
CBHW041748220926
51847CB00013B/142
* 9 7 8 1 9 4 7 4 8 2 1 2 8 *